Microbiology: A Very Short Introduction

VERY SHORT INTRODUCTIONS are for anyone wanting a stimulating and accessible way in to a new subject. They are written by experts, and have been translated into more than 40 different languages.

The Series began in 1995, and now covers a wide variety of topics in every discipline. The VSI library now contains over 350 volumes—a Very Short Introduction to everything from Psychology and Philosophy of Science to American History and Relativity—and continues to grow in every subject area.

Very Short Introductions available now:

For more information visit our website

www.oup.com/vsi/

Nicholas P. Money

MICROBIOLOGY

A Very Short Introduction

OXFORD
UNIVERSITY PRESS

OXFORD
UNIVERSITY PRESS

Great Clarendon Street, Oxford, ox2 6DP,
United Kingdom

Oxford University Press is a department of the University of Oxford.
It furthers the University's objective of excellence in research, scholarship,
and education by publishing worldwide. Oxford is a registered trade mark of
Oxford University Press in the UK and in certain other countries

First edition published in 2014

Published in the United States of America by Oxford University Press
198 Madison Avenue, New York, NY 10016, United States of America

British Library Cataloguing in Publication Data
Data available

Library of Congress Control Number: 2014943834

ISBN 978-0-19-968168-6

Printed and bound by
CPI Group (UK) Ltd, Croydon, CR0 4YY

The manufacturer's authorised representative in the EU for product
safety is Oxford University Press España S.A. of el Parque Empresarial
San Fernando de Henares, Avenida de Castilla,
2 – 28830 Madrid (www.oup.es/en).

Contents

List of illustrations

Microbiology

Chapter 1
Microbial diversity

Earth is dominated by microorganisms. It can be difficult to appreciate this fact for the obvious reason that these forms of life are invisible to the unaided eye. We see plants and animals and interact with them in a deliberate fashion, and for most of human history we had no proof that anything smaller than insects existed. The Roman philosopher Lucretius edged close to the truth with his conjecture that 'certain minute creatures...enter the body through the mouth and nose and cause serious diseases'. His musings began to make sense after the invention of the microscope in the 1600s. The numbers of microbes are staggering. Tens of millions of bacteria live in a pinch of soil; a drop of seawater contains 500,000 bacteria and tens of millions of viruses; the air is filled with microscopic fungal spores, and a hundred trillion bacteria swarm inside the human gut. Every macroscopic organism and every inanimate surface is coated with microbes; microbes grow around volcanoes and hydrothermal vents; they live in blocks of sea ice, in the deepest oceans, and thrive in ancient sediment on the seafloor. Microbiology is the scientific study of these smallest forms of life. It concerns the biology of the bacteria, archaea, fungi, and an astonishing variety of unicellular organisms called protists. Microbiologists also study viruses, whose structure is far simpler than any kind of cell.

The majority of macroscopic organisms rely upon the energy harvested from the sun by photosynthesis: they are plants, they eat plants, or they consume animals that eat plants. Microorganisms show a greater range of metabolic lifestyle. The most familiar species act as decomposers, recycling the substance of dead plants and animals. Other microbes are photosynthetic. These include the cyanobacteria and a variety of protists that we call algae. In addition to decomposers and photosynthetic microbes, diverse bacteria and archaea are powered by metabolic processes fuelled by hydrogen gas, sulfur, and simple molecules including ammonia and methane. These chemical pathways enable microorganisms to support entire ecosystems in locations of perpetual darkness. The biochemical virtuosity of bacteria and archaea on Earth has encouraged astrobiologists to speculate about microbial life in a subsurface ocean on Jupiter's moon Europa and in a methane-fuelled ecosystem on Saturn's moon Titan.

Experiments in optics at the beginning of the 17th century allowed European scientists, including Galileo, to develop the first microscopes shortly after the invention of the telescope. The earliest microscopic observations were made on insects and the first illustrations of microorganisms were published in 1665 by Robert Hooke, who described the spore-producing structures of fungi. Hooke's contemporary Anton van Leeuwenhoek went much further in his investigations on the microbial world, being the first to describe bacteria, including large, crescent-shaped cells scraped from his teeth, a variety of protists, and yeast from beer. Despite important microscopic investigations by a handful of clever scientists in the 1700s, little progress was made in microbiology until the next century when Louis Pasteur demonstrated that sterilized broth remained sterile as long as it was isolated from airborne microbes. This rigorous experimental work in the 1860s disproved classical ideas about the spontaneous generation of organisms. Later, Pasteur developed vaccines against anthrax (caused by the bacterium *Bacillus anthracis*) and rabies (caused by a virus). Using mice in his experiments, Robert Koch identified the anthrax bacterium in the 1870s and designed a systematic method for identifying the cause of any infectious disease. This method, called Koch's Postulates, requires the investigator to identify the disease-causing organism in a diseased animal, grow this 'germ' in a pure culture, use this culture to infect a healthy animal, and isolate the same organism from the experimental host.

Even after the invention of the microscope, however, the classical Aristotelian division of life into animals and plants was unchallenged until Ernst Haeckel created a third category for unicellular species, named 'protista', in the 1860s. Today, we recognize three primary groups of organisms: Bacteria, Archaea, and Eukarya (Figure 1). Informal names of these groups—bacteria, archaea, and eukaryotes—are used throughout the rest of this book. Microbiologists study microscopic organisms belonging to

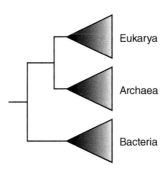

1. Three primary groupings, or domains, of organisms displayed in the form of an evolutionary 'tree' in which genetic modification over the course of billions of years is indicated from left to right. The diagram shows that Bacteria, Archaea, and Eukarya evolved from a common ancestor, and that the Archaea and Eukarya are more closely related to one another than either group is connected to the Bacteria. Genetic comparisons suggest that the Eukarya may have evolved from a group of Archaea. According to this research it makes sense to condense all organisms into a pair of domains, the Bacteria and Archaea

all three groups. All bacteria and archaea are microscopic; soil amoebae, diatoms, dinoflagellates, and single-celled green algae are examples of microscopic eukaryotes.

There is a crucial distinction between bacteria and archaea, which are prokaryotes, and the eukaryotic microbes (Figure 2). Genes of prokaryotes are organized in the form of a single circular chromosome situated in the fluid interior of the cell. This chromosome constitutes the genome of prokaryotes. The genome is defined as the entire collection of hereditary information in the cell. Genomes, microbial and otherwise, comprise genes that encode proteins, intervening sequences that regulate gene expression, and non-coding sequences that do not specify proteins and used to be called 'junk DNA'. (We know now that a great deal of the non-coding DNA performs important biological functions.) Eukaryotes tend to carry more genetic information than prokaryotes. Most of the genome of a eukaryote cell is encoded in multiple chromosomes surrounded by an envelope of membranes that defines the nucleus. DNA is also found in the form of circular chromosomes inside the mitochondria and

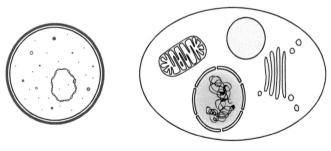

Prokaryote   Eukaryote

2. Diagram showing the relatively simple cell structure of prokaryotes (bacteria and archaea) contrasted with the more complex makeup of eukaryotes. The single chromosome of bacteria and archaea is situated within the cytoplasm. The multiple chromosomes of eukaryotes are housed within the nucleus

chloroplasts of eukaryotic cells. Mitochondria are organelles that supply eukaryote cells with energy. Chloroplasts are organelles that perform photosynthesis in plants and algae. Both types of organelle evolved from bacterial cells that were absorbed by the ancestors of today's eukaryotes. This process of incorporation is called endosymbiosis.

Structural and functional characteristics like cell shape and metabolic activity have been used to identify some of the groupings of bacteria since Pasteur's time. Cells of the syphilis bacterium (*Treponema pallidum*), for example, are coiled like a corkscrew. Their unusual shape is evident in samples of infected tissue from syphilis patients and is characteristic of related bacteria that we classify as Spirochetes. These structural features are a less reliable guide to relatedness in other cases and most rod-shaped bacteria and archaea look the same under the microscope. The Gram-staining technique, developed in the 19th century, is another guide to identification. This is described as a differential stain because it colours bacteria with thick cell walls purple (positive) and thin-walled bacteria pink (negative). Staining reactions are a useful diagnostic tool, allowing a medical technician to narrow the list of possible bacteria collected from a throat swab. But the Gram stain is a poor guide to the relatedness of species. For this reason, microscopic methods have been largely superseded by genetic techniques for developing modern classification schemes that reflect evolutionary kinship.

Research on evolutionary relatedness, known as molecular phylogenetic analysis, relies upon comparisons between the DNA sequences of different species. For bacteria, a gene that encodes part of a cell structure called the ribosome is crucial for the identification of species. Ribosomes are molecular machines that produce proteins. In general, if the sequence of this 16S ribosomal RNA (rRNA) gene of different bacterial isolates, or strains, differs by 3 per cent or less, these microorganisms are regarded as members of the same species. This is not a perfect method, and probably underestimates the number of species, but it is very useful for identifying bacteria from environmental samples. Comparisons between 16S rRNA genes are used to construct phylogenetic trees that reveal evolutionary relationships between different species and links between groups of bacteria (Figure 3). Analysis of other genes is essential for discriminating between strains within a single species. Comparisons of whole genomes have also been used with spectacular success to examine the details of bacterial evolution.

Specialists who study bacterial taxonomy have catalogued more than 11,000 species of bacteria. This list is biased toward bacteria of medical significance and those that can be grown in culture easily. Experiments in which bacterial genes have been sequenced without growing the cells in the laboratory show that a teaspoon of soil can contain thousands of unidentified species. Our bodies are home to an incredible range of microorganisms: molecular analysis has revealed 2,368 'species' of bacteria that live in the human navel! These results have encouraged researchers to suggest that there may be tens or even hundreds of millions of species of bacteria. Undercounting of bacteria, and other microorganisms,

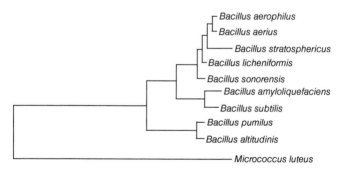

3. Phylogenetic tree showing relationships between species of *Bacillus* (bacteria) based upon comparisons between the sequences of their ribosomal RNA genes. The outgroup chosen as a reference for comparing the *Bacillus* species is the bacterium *Micrococcus luteus*

is an important theme in the contemporary struggle to develop objective measurements of biodiversity.

With these uncertainties in mind, eighty or more sub-groups, or phyla, of bacteria have been named by microbiologists. Proteobacteria constitute the largest bacterial phylum and they incorporate all manner of cell shapes and mechanisms of energy generation. Proteobacteria include *Escherichia coli*, a gut bacterium that has been studied as a model experimental organism by geneticists since the 1940s; pathogenic bacteria that cause typhoid (*Salmonella enterica*) and cholera (*Vibrio cholerae*); nitrogen-fixing bacteria; purple photosynthetic bacteria; stalked bacteria, and myxobacteria that form beautiful multicellular fruit bodies. Proteobacteria are Gram-negative bacteria whose cells are surrounded by a pair of lipid membranes, called the inner and the outer membrane, separated by a relatively thin cell wall. Bacterial walls are built from a polymer called peptidoglycan. Peptidoglycan is built from chains of alternating pairs of the amino sugar molecules N-acetylglucosamine and N-acetylmuramic acid cross-linked by peptides. Lysozyme is an antibacterial enzyme found in tears, human milk, and mucus, which kills bacteria by disrupting the bond between the amino sugars in the peptidoglycan wall. Penicillin also targets the bacterial wall and works by inhibiting peptidoglycan synthesis.

The peptidoglycan wall is much thinner in proteobacteria and other Gram-negative groups than the cell wall of Gram-positive bacteria including a group called the firmicutes. *Clostridium* and *Bacillus* are firmicutes that produce thick-walled cells called endospores. Gram-positive bacteria lack the outer membrane characteristic of Gram-negative species. Bacteria in a third phylum, the mollicutes, have no peptidoglycan wall. These are better known as the mycoplasmas and several of them cause diseases in mammals. Because they lack cell walls they are resistant to lysozyme and to antibiotics that target the peptidoglucan polymer. Some mycoplasma cells are less than 0.2 micrometres (μm) in diameter. This compares with an average size of 1 μm for walled bacteria. There are a few examples of gigantic bacteria, including a sulfur-oxidizing bacterium, *Thiomargarita namibiensis*, whose cells have a diameter of 750 μm or 0.75 millimetres (mm).

Actinobacteria are characterized by a filamentous growth form. They include the pathogens that cause tuberculosis (*Mycobacterium tuberculosis*) and leprosy, or Hansen's disease (*Mycobacterium leprae*), and *Streptomyces* species that produce streptomycin, tetracycline, and other antibiotics. *Streptomyces* produces colonies of filaments and aerial branches that fragment into chains of spores. Filamentous growth forms are also produced by some of the cyanobacteria (Figure 4). Photosynthesis by marine cyanobacteria is a major component of global carbon fixation. Many cyanobacteria absorb atmospheric nitrogen and form ammonia and other compounds, playing a critical role in the

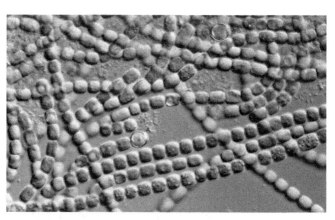

4. Cell filaments of a photosynthetic cyanobacterium. The pair of circular thick-walled cells within the filaments are called heterocysts and function in nitrogen fixation. The chloroplasts of photosynthetic eukaryotes evolved from cyanobacterial cells that were absorbed by the ancestors of eukaryotic algae more than one billion years ago

nitrogen cycle. Some cyanobacteria live inside root nodules in legumes and others grow with fungi to form lichens. These relationships are called mutualistic symbioses, or mutualisms for short.

Phylum Deinococcus-Thermus is a small group which includes bacteria with remarkable tolerance to ionizing radiation and heat. Humans are killed with a single dose of 5 Gy, or grays, of ionizing radiation, equal to 5 joules of energy per kilogram of body mass. *Deinococcus radiourans* can withstand 15,000 Gy and has been called, 'Conan the Bacterium'. Its relative, *Thermus aquaticus*, was isolated in Lower Geyser Basin in the Yellowstone National Park in the United States in the 1960s and thrives at 70 °C (158 °F). An enzyme in this species called *Taq* polymerase was the original enzyme used for the polymerase chain reaction (PCR). The ability of Taq polymerase to copy DNA at high temperatures revolutionized molecular genetic research, provided forensic investigations with a scientific foundation, and found a wealth of applications in modern medicine.

Many bacteria are motile and employ the rotation of flagella to push or to pull them through fluid environments. Bacterial flagella are marvels of evolutionary engineering (Figure 5). The filament that trails from the cell surface is a hollow tube made from 30,000 subunits of a single protein called flagellin. This is attached via a bend, called the hook, to the basal body of the flagellum. The basal body consists of a rod that passes through a series of protein rings in the outer membrane (of Gram-negative bacteria), the peptidoglycan wall, and the inner membrane where it is connected to a rotor. Additional proteins in the inner membrane (Mot proteins) operate as channels for the transmission of protons (hydrogen ions symbolized as H^+). Ion flow through these channels causes conformational changes in the Mot proteins that spin the rotor. The whole assembly is a reversible nanoscale electric motor that twirls the filament at 100 revolutions per minute.

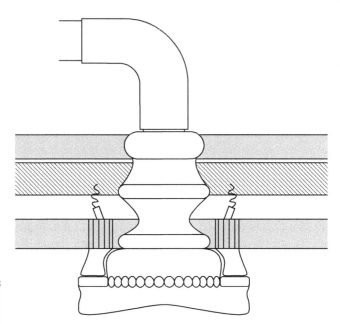

5. Complex structure of the bacterial flagellum with protein components organized as rings in the cytoplasm (cell interior), inner membrane, peptidoglycan wall, and outer membrane of Gram-negative bacteria. Proteins in the inner membrane, which are tethered to the wall, spin the assembly at 100 revolutions per second

Bacteria can be propelled by a single flagellum (polar arrangement), by tufts of the motors concentrated at one end or both ends of the cell (lophotrichous and amphitrichous), or by numerous flagella anchored at multiple points on the surface (peritrichous). Bacteria are so small that their motion is dominated by the viscosity of their surroundings; they have zero inertia. Cells are pushed or pulled through water by their beating flagella, and stop dead the moment the flagella motors switch off. A typical swimming speed is 25 μm per second, which is comparable, in relation to the size of the organism, to the top speed of a cheetah. Filamentous cyanobacteria and myxobacteria do not have flagella and use gliding mechanisms involving slime extrusion or the extension of surface proteins to move over surfaces or to slide past one another.

In contrast to the large number of subgroups of bacteria, most of the 500 named species of archaea are divided into just two phyla: the euryarchaeota and the crenarchaeota. Together, these microorganisms inhabit extreme environments including hot springs, hydrothermal vents on the ocean floor, salt-saturated pools, and highly acidic and highly alkaline habitats. Archaea also grow in the open ocean where they are a major component of the deep water plankton and fertilize the water by converting ammonia (NH_3) to nitrite (NO_2^-). This ecological diversity is enabled by a range of physiological mechanisms that allow archaea to operate in aerobic and anaerobic habitats and power themselves with the chemical energy in hydrogen molecules (H_2), sulfur atoms (S^0), and compounds containing iron. Methanogens capture or 'fix' CO_2 using H_2 as an energy source. Some of the carbon fixed by these species is used to synthesize their cellular materials and the rest is released as methane gas (CH_4). Methane-generating archaea populate the digestive systems of termites and ruminants, and are an important part of the community of microbes, or microbiome, in the human gut. Although there are no photosynthetic archaea, the haloarchaea, which live in hypersaline pools, use a light-absorbing pigment called bacteriorhodopsin to power themselves at low oxygen levels.

Like bacteria, archaea have tiny cells and their genomes are encoded in single circular chromosomes. Archaea do not have peptidoglycan walls or outer membranes. The most common type of wall in archaea is called the S-layer (S stands for surface). This is formed from interlocked protein or glycoprotein molecules and looks like a tiled floor when it is viewed with an electron microscope. Some methane-producing archaea have a pseudomurein wall, which, like bacterial peptidoglycan, contains chains of amino sugar molecules cross-linked by peptides. Common cell shapes among archaea are rods, cocci (rounded cells), and filaments, but the cells of some haloarchaea grow as flattened squares and triangles. *Thermoplasma* and *Ferroplasma* are archaea that lack cell walls. These microorganisms live in hot acidic soils. Some of the wall-less archaea are only 0.2 µm in diameter, resembling the bacterial mycoplasmas. Motile archaea are equipped with tufts of flagella whose molecular structure is completely different from the flagella of bacteria.

Bacteria and archaea have considerably smaller genomes than eukaryotes. The circular chromosome of the bacterium *Escherichia coli* (strain K-12) contains 4,288 genes that encode proteins. Myxobacteria have the largest bacterial genomes that encode more than 9,000 genes. This genetic complexity is related to the life cycles of the myxobacteria entailing the formation of multicellular fruit bodies. The smallest bacterial genomes are found in species that are symbionts that live inside the cells of insects. Archaeal genomes encode up to 4,300 genes. For comparison, consider that the twenty-three pairs of human chromosomes encode more than 20,000 genes.

Microscopic eukaryotes do not utilize the range of metabolic mechanisms found in the bacteria and archaea. Algae (and plants) manufacture food by the kind of photosynthesis used by cyanobacteria, and all other eukaryotes meet their energetic needs by digesting other organisms or their waste products. Lacking metabolic variety, other measures of eukaryote diversity are evident at the genetic level and this is manifested in a tremendous range of structural and morphological characteristics. There are eight supergroupings of eukaryotes (Figure 6); all of them include single-celled organisms and five are entirely microbial. Amoebozoans include *Amoeba proteus*, the iconic simple form of pond life that moves by extending its finger-like pseudopodia at one end of the cell and retracting its posterior cytoplasm. It feeds by engulfing bacteria and other organisms through the process of phagocytosis. Phagocytosis is used by eukaryote cells to absorb

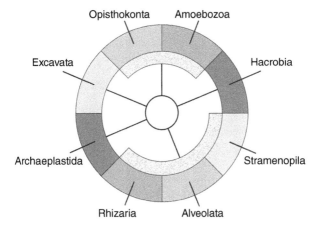

6. The wheel of eukaryote life. Eight supergroups arranged around circumference are linked by spokes to the central hub symbolizing the ancestral eukaryote. Common prototypes that unite some of the groups are indicated by the intermediate segments

particles, including bacteria, which they digest inside their cytoplasm. This serves as a feeding mechanism and is used by cells in the human immune system to destroy pathogens. Certain kinds of slime mould are amoebozoans too. The slime mould *Dictyostelium discoideum* is described as a species of social amoeba whose individual cells stream together to produce a stalked fruit body with spores at its tip.

Hacrobia include cryptomonad algae whose cells contain four separate genomes (Figure 7). Ancestors of photosynthetic cryptomonads fed themselves by phagocytosis and acquired their chloroplasts by engulfing a red algal cell that resisted digestion. This is an example of the evolutionary mechanism of endosymbiosis introduced earlier. Cryptomonad chloroplasts contain a cell structure called the nucleomorph which is a remnant of the nucleus of the original red algal partner (I), plus a circular chromosome of cyanobacterial origin (II). The cyanobacterial

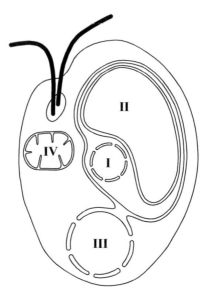

7. Complex structure of a cryptomonad alga produced by fusion of an amoeboid cell with a photosynthetic red alga. Evidence of this process is found in a structure called the nucleomorph trapped between the membranes of the chloroplast (I). The nucleomorph is an accessory genome derived from the nucleus of the red algal partner

chromosome is evidence that the ancestor of the red alga engulfed a bacterium that was coopted as its chloroplast. Played out over many millions of years, a eukaryote containing a nucleus absorbed a cyanobacterium, which became its chloroplast, and this chimeric cell (a red alga) was absorbed by another eukaryote with its own nucleus (III) and was transformed into *its* chloroplast. The fourth genome in the cryptomonad is found in the mitochondria (IV). Mitochondrial chromosomes are bacterial in nature and this acquisition was performed by the ancestor of all eukaryotes.

Diatoms are examples of photosynthetic stramenopiles. Marine diatoms, together with planktonic cyanobacteria, absorb as much carbon dioxide as all of the land plants and produce half of the oxygen in our atmosphere. Water moulds are a second kind of stramenopile. They produce swimming cells, called zoospores, and resemble fungi by forming colonies of filamentous cells that penetrate their food sources. *Phytophthora infestans* is the species that caused the potato blight epidemic that devastated the Irish population in the 1840s. Giant kelps, whose fronds can extend to a length of 50 metres, are the largest of the stramenopiles. Alveolates and rhizarians are exclusively unicellular. Alveolates include photosynthetic and non-photosynthetic dinoflagellates that are exceedingly common planktonic microbes in the freshwater and marine environment; radiolarians and foraminiferans are groups of marine rhizaria.

Many of the green algae are microorganisms and are related to the plants with which they are combined in the archaeplastid supergroup. Microscopic green algae include tiny floating spheres, *Chlamydomonas* and other cells that form flagellated cells, gorgeous star-shaped desmids, and the multicelled colonies of *Volvox*. Excavates include euglenoid algae and the intestinal parasite *Giardia* spread by contaminated water.

Opisthokonts are the eighth supergroup of eukaryotes that incorporates the fungi and animals. Genetic similarities between these seemingly disparate forms of life is undeniable and we share the common signature of flagellate cells. Motile zoospores of aquatic fungi swim using single flagella, and animals produce many types of flagellated cells (cells are called ciliated when they have multiple flagella). More than 70,000 species of fungi have been described and the total number may exceed one million. In addition to the motile aquatic species, fungi grow as single-celled yeasts that reproduce by forming buds, and as colonies of filamentous cells that can spread over huge territories. Mushroom development by basidiomycete fungi redistributes the subterranean biomass accumulated during the feeding phase of the colony into aboveground fruit bodies for the purpose of spore formation and dispersal. The majority of fungi do not produce any macroscopic structures and form spores directly on the surface of their colonies.

A striking conclusion from the supergroup view of eukaryotes is that so little of the diversity of life is organized in the form of big organisms. Plants and animals are the macrobiological exceptions to the preponderance of microorganisms in the biosphere. Plants are defined as the large organisms within the archaeplastids. And although most animals are macroscopic, soil nematode worms, planktonic crustaceans, rotifers, and dust mites are examples of microscopic or near-microscopic animals. Organization of these supergroups is based on scrutiny of a DNA sequence called the ITS region. The ITS region encodes three different components of the eukaryote ribosome (the protein-producing machine) and spans intervening sequences of DNA called internal transcribed spacers. This has shown that the opisthokonts and amoebozoans share a common ancestor, and that the stramenopiles, alveolates, and rhizarians are unified through another root to the first eukaryotes. Each of the remaining trio of supergroups has its own unique genealogical connection to this progenitor. Besides its value in evolutionary research, amplification of eukaryote ribosomal RNA sequences is useful for environmental sampling where it complements the identification of prokaryotes from their 16S ribosomal RNA sequences. Like prokaryotes, a lot of the eukaryotic microorganisms cannot be cultured and it is very difficult to discriminate between species using a microscope. The bias favouring animals, fungi, and plants in surveys of biodiversity is evident from genetic analysis of environmental samples that reveals an incredible variety of eukaryotic and prokaryotic microbes. Amplification of genes from samples of soil, seawater, freshwater, and even chlorinated drinking water led to the discovery of a new phylum of unicellular fungi in 2011, given the formal name Cryptomycota. Genetic diversity in this enigmatic collection of species appears to be greater than all of the fungi known previously.

Unlike eukaryote metabolism, eukaryote structure is highly variable. Beginning with the cell surface, some of the amoebozoans create protective shells by organizing mineral particles and other debris on their surface, diatoms secrete beautiful silica glass walls called frustules, and the cell walls of fungi are strengthened with microfibrils of chitin. The interior of the eukaryote cell contains the nucleus, a complex secretory apparatus called the endomembrane system, mitochondria, vacuoles, and other organelles surrounded by membranes. The flagella of eukaryotes are more complex than those of bacteria and archaea. They project from the cell within a sheath of cell membrane and contain a series of long filaments called microtubules that slide up and down like pistons causing the flagellum to bend. Movement of the microtubules is driven by a motor protein called dynein. Flagellar microtubules are rooted inside the cell where they connect with a larger cytoskeleton of microtubules and a web of actin microfilaments. This internal cytoskeleton is involved in growth and development, controls chromosome movement and cell division, and forms a guidance system for materials travelling to and from the cell surface.

In general, eukaryote cells are larger than cells of bacteria and archaea, ranging from 10 to 100 μm (or 0.1 mm). Eukaryote cells can be studied even at low magnifications using a light microscope. Bacteria can appear as a background haze at these magnifications. For this reason, microbiologists often use a 100× objective lens, coupled with 10× eyepiece lenses, to count bacteria, track their motion, and record the results of Gram staining. At this total magnification of 1,000× structures inside bacteria are still invisible, but variations in cell shape are clear. The tiniest cells produced by mycoplasmas (0.2 μm or 200 nanometres (nm) are at the edge of detection using a conventional light microscope. This is because they are smaller than half the wavelength of visible light and do not produce the separated patterns of diffracted light that enable us to perceive discrete objects (the shortest visible wavelength at the violet end of the spectrum is 390 nm). Viruses range in size from 20 to 600 nm, making most of them invisible under the light microscope. The much shorter wavelength of the electron beam used in electron microscopes provides more than one million-fold magnification, clarifying structures with a diameter of 0.1 nm and revealing details of viral structure.

Viruses are not regarded as living organisms because they lack their own 'onboard' metabolism and have to infect living cells to replicate. Other than the initial infection process, all of the biological activity of a virus occurs inside its host. 'Molecular organism' is a term used by some scientists to describe viruses and is useful for distinguishing them from prokaryotes and eukaryotes, or 'cellular organisms'. Viral DNA or RNA is packaged inside a protein shell called the capsid (Figure 8). Some virus capsids are surrounded by a lipid membrane or envelope. This simple combination of structures has been adapted to enable viruses to infect every kind of cellular organism. DNA and RNA bacteriophages attack bacteria; filamentous and spindle-shaped DNA viruses infect archaea; viruses target unicellular eukaryotes in all of the supergroups, and fungi, animals, and plants. Genome sizes vary a lot. The single-stranded DNA molecule of a circovirus that infects animals encodes only two proteins: one protein is the subunit that is assembled into the capsid and the other drives the replication of the circovirus DNA within the host cell. At the other extreme of viral genome sizes are mimivirus genomes that encode more than 1,000 proteins. Mimiviruses are classified as nucleocytoplasmic large DNA viruses. These complex viruses have a lipid layer inside the capsid and contain enzymes that catalyse DNA synthesis and transcription. They even produce enzymes that control several metabolic reactions, which makes it difficult to say whether they are molecular or cellular organisms.

Some viral infections result in the incorporation of viral genes into the genome of the host organism. This dormant form of the virus is called proviral latency. Viruses can also carry genetic information between hosts, driving horizontal transfer of genes.

8. The structure of a virus showing outermost lipid envelope studded with glycoproteins, interior protein capsid, and genome in the centre of the particle encoded in DNA or RNA strands

These processes are exceedingly important in evolution, altering the genetics of cellular organisms and allowing infected hosts to transmit genes of viral origin through their own reproductive mechanisms. Human DNA, for example, may contain 100,000 fragments of genes from endogenous retroviruses, constituting 8 per cent of our genome. Viruses are also highly significant players in global ecology. One millilitre of North Atlantic seawater contains fifteen million bacteriophages that destroy an estimated 40 per cent of the planktonic cyanobacteria every day. Viruses are everywhere and much of the genetic diversity in the biosphere is carried within them.

Chapter 2
How microbes operate

In this chapter we will consider the mechanisms that sustain prokaryotic and eukaryotic microorganisms. All active cells must be supplied with water and an energy source. Absorption of water is essential, even in extremely dry or salty habitats, because the enzymes that catalyse biochemical reactions in cells do not work unless they are hydrated. Sunlight powers the metabolism of photosynthetic microbes and others glean chemical energy from a plenitude of terrestrial sources. Extremes in temperature, acidity, and other environmental variables place additional constraints upon microbial life, but bacteria, archaea, and eukaryotic microorganisms thrive in most places where liquid water is available.

Considerations about water availability are crucial for understanding how microbes work. When a bacterial cell is exposed to dry air, it loses water by evaporation. This happens because water molecules diffuse from the cell, where they are concentrated, into the air where they are less concentrated, according to the thermodynamic imperative that favours disorder or higher entropy. When a bacterial cell is submerged in water, the flux is reversed: the concentration of biomolecules and ions dissolved in the bacterial cytoplasm (collectively called solutes) supports the absorption of water through the cell membrane. This is an example of osmosis. Hydration and dehydration of cells is determined by the difference in the concentration of solutes in the cytoplasm and in the external environment. As long as the cell is more salty or sugary than its surroundings it will remain hydrated. This is true for an aquatic fungus growing in the cold water of a mountain stream, bacteria in the colon, and populations of archaea living in brine pools saturated with salt.

As cells absorb water they expand and their cytoplasm is diluted. Expansion is limited in most microorganisms by the development of hydrostatic pressure or 'turgor' as the cell membrane is pushed against the inner surface of the cell wall. Archaea in brine pools have little or no turgor because they are only slightly saltier than their surroundings. Other microbes have average pressures of 1–5 atmospheres (up to 500 kilopascals). Turgor affects cell shape and is used by filamentous fungi to push themselves through solid materials in search of food. This pressure-aided invasive growth is an important infection mechanism used by fungal parasites to penetrate plants.

The cell membrane (inner membrane of Gram-negative bacteria) is the semipermeable barrier that permits water influx while retaining cytoplasmic solutes. It is composed of lipid molecules, organized in two layers, and membrane proteins that perform a variety of functions. Some water diffuses through the lipid part of the membrane and faster conduction of water occurs through proteins called aquaporins. Specific solutes dissolved in water move through other kinds of transmembrane proteins that thread from one side of the membrane to the other (Figure 9). Transmembrane proteins called channels allow ions to flow into or out of the cell according to the direction of their concentration gradients. Potassium ions (K^+), for example, pass through K^+-specific ion channels. Ion transit through channels is referred to as facilitated diffusion. Solutes are also transported *against* their concentration gradients by protein pumps powered by chemical energy or sunlight. Other transport proteins convey solutes against their concentration gradients by linking their

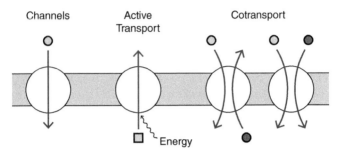

9. Transmembrane proteins transport ions and molecules between cells and the surrounding environment

activity to the passive movement of another solute down its concentration gradient. The combination of the permeability barrier of the membrane and operation of its transport proteins enables the cell to maintain a chemical environment conducive to its metabolic activity.

Cells of unicellular and multicellular organisms are powered by the oxidation and reduction of inorganic and organic compounds. Oxidation reactions remove electrons from atoms, ions, or molecules and reduction reactions add electrons. In the biological context, energized electrons are released by oxidation and these are used to reduce, or fix, carbon dioxide to produce sugars and other organic compounds. This loop of redox reactions occurs in two types of microorganism: photolithotrophs and chemolithotrophs (Figure 10). Organisms in both groups are called autotrophs because they manufacture their own food. Photolithotrophic (photosynthetic) bacteria and eukaryotic algae use sunlight to energize electrons obtained from water or hydrogen sulfide and use this reducing power to absorb carbon dioxide and create sugars and other biological molecules. 'Solar powered' is a simpler term for this metabolic category than photolithotroph. Chemolithotrophic bacteria and archaea obtain their reducing power by harvesting electrons from sulfur (S^0), ferrous iron (Fe^{2+}), and inorganic compounds including hydrogen (H_2), hydrogen sulfide (H_2S),

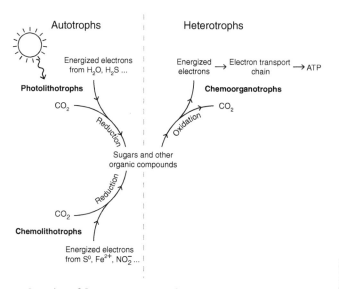

10. Overview of the energy sources of autotrophs (that make their own food) and heterotrophs (that consume biological molecules produced by other organisms). Autotrophs oxidize some of the organic compounds that they generate to meet their energy needs and utilize electron transport chains to produce ATP (as shown on right-hand side of diagram for the chemoorganotrophs)

ammonia (NH_3, which dissolves as ammonium ions, NH_4^+), and nitrite (NO_2^-). Chemolithotrophs are 'mineral powered', with the exception of species that use methane (methanotrophs) and other simple organic compounds rather than inorganic sources of electrons.

Every other form of life is dependent upon solar- and mineral-powered organisms because they feed upon, or oxidize, organic compounds produced by autotrophs. These organisms are chemoorganotrophs, also known as heterotrophs. They include fungi and bacteria that decompose (oxidize) organic materials and consume (reduce) oxygen, and denitrifying bacteria that oxidize organic compounds in the absence of oxygen and reduce nitrate

(NO_3^-) to nitrogen gas (N_2). Similar reactions are performed by diverse heterotrophic bacteria and archaea that reduce sulfate (SO_4^{2-}), ferric iron (Fe^{3+}), and manganese (in the form MnO_2). (We are heterotrophs too, oxidizing sugars and reducing oxygen in our respiratory reactions.)

Modified versions of these metabolic strategies are scattered among the bacteria, archaea, and fungi. These include light-absorbing photoheterotrophs that circumvent carbon dioxide fixation and use organic compounds from the environment as their carbon source and mixotrophs that combine different feeding mechanisms to take advantage of changing environmental conditions. All of the ways of making a living evolved first among microorganisms.

Closer examination of these categories of metabolism shows how microorganisms make use of redox chemistry to power their activities. Primary production by photosynthetic cyanobacteria begins with the absorption of light by chlorophyll and other pigment molecules integrated into membranes called thylakoids. Thylakoids are arranged as concentric sheets inside the cell (Figure 11). Chlorophyll molecules are organized into multiple copies of two kinds of assemblage, called photosystems, which boost the energy level of electrons when they absorb sunlight. One of the photosystems splits water molecules, releases oxygen, and passes the energized electrons through a series of carrier molecules located in the membranes. The carrier molecules form an electron transport chain (Figure 12). When the electrons reach the second photosystem the absorption of sunlight raises their energy level again and the re-energized electrons are transferred to a molecule called NADPH. NADPH serves as a portable source of reducing power that is used to fix CO_2 to produce the biomolecules that form the structure of the cell.

Energy is tapped from the electrons as they are transmitted through the electron transport chain between the two photosystems and this is used to pump protons (H^+) across the thylakoid membrane.

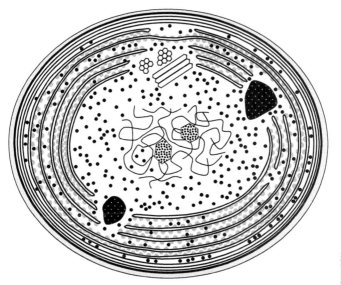

11. The photosynthetic pigments of cyanobacteria are organized in membranes called thylakoids inside the cell of a cyanobacterium. These are not separate organelles like the chloroplasts of eukaryotic algae and plants

Charge separation across membranes is central to the generation of energy in all organisms. In the case of the photosynthetic cyanobacteria, the resulting concentration gradient of protons is used to produce adenosine triphosphate (ATP). ATP serves as an energy source for biochemical reactions.

Oxygenic photosynthesis—that releases oxygen from the reaction that splits water molecules—is also performed by plants and varied eukaryotic microorganisms, including cryptomonads, dinoflagellates, diatoms, euglenoids, and red, green, and brown algae. These groups of photosynthetic algae are specialized at absorbing light of different wavelengths and intensities. A remarkable species of red alga lives in apparent darkness at a depth of 268 metres in the Bahamas by absorbing the exceedingly

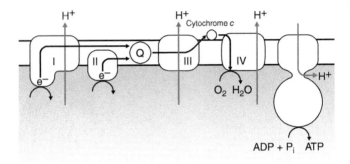

12. Electron transport chains harness energy within the cell membranes of bacteria and archaea and inside the mitochondria and chloroplasts of eukaryotes

weak blue light. In all of the photosynthetic eukaryotes, the thylakoid membranes bearing the photosystems are situated inside chloroplasts. As we saw in Chapter 1, chloroplasts were derived, via endosymbiosis, from cyanobacteria or from photosynthetic eukaryotes whose ancestors had absorbed cyanobacteria.

Photosynthetic purple and green sulfur bacteria use hydrogen sulfide (H_2S) rather than water as their source of electrons, capturing CO_2 without releasing oxygen (this is non-oxygenic photosynthesis). Light-absorbing pigments of the sulfur bacteria are arrayed on stacks of internal membranes connected to the cell membrane and in blobs called chlorosomes. Unlike cyanobacteria, sulfur bacteria use a single type of photosystem. When this absorbs light, energized electrons are passed to an electron transport chain that produces a proton gradient and returns the depleted electrons to the same kind of photosystem. This cyclic flow of electrons in non-oxygenic photosynthesis differs from the noncyclic flow between two kinds of photosystem in oxygenic photosynthesis. Non-oxygenic photosynthesis also occurs in heliobacteria, acidobacteria, and purple and green non-sulfur bacteria. A species of green sulfur bacterium accomplishes the astonishing feat of conducting photosynthesis in a black smoker hydrothermal vent at a depth of more than 2 kilometres beneath the surface of the ocean. It does so by absorbing photons from flashes of geothermal light.

Chemolithotrophic or mineral-powered bacteria and archaea use enzymes to oxidize their inorganic fuels and funnel the resulting electrons through electron transport chains that function like the series of carrier molecules used in photosynthesis. The electron transport chains of chemolithotrophs are located in the cell membrane and the flow of electrons is used to pump protons (H^+) and generate ATP. At the end of the electron transport chain, the spent electrons are transferred to oxygen or an alternative electron acceptor. This flow of electrons is also used to produce a portable source of reducing power in the form of NADH. Like the NADPH produced in photosynthesis, NADH is used to reduce CO_2 to produce the biomolecules that form the structure of the cell. Some of the chemolithotrophs can operate as heterotrophs when sugars and other organic compounds are available in their environment. This physiological flexibility is an example of mixotrophy.

Sites of volcanic activity and environments polluted by mining and agricultural activities are particularly favourable for the growth of chemolithotrophs because they are enriched in their inorganic fuels. Hydrogen-oxidizing bacteria and archaea occupy hot springs and hydrothermal vents on the ocean floor; sulfur-oxidizing bacteria thrive in the same sorts of habitat, and iron-oxidizing bacteria are abundant in the acidic water flowing from abandoned coal mines and other toxic waste sites. By oxidizing ammonia to nitrite, and nitrite to nitrate (NO_3^-), nitrifying bacteria act as natural fertilizers in soils and lakes, supporting the growth of plants and photosynthetic microbes. Like other chemolithotrophs, nitrifying bacteria manufacture their biomolecules by fixing CO_2. In marine environments, ammonia-oxidizing archaea produce nitrite and 'anammox' bacteria oxidize ammonia under anaerobic conditions releasing nitrogen gas (N_2).

Nitrogen fixation by prokaryotes is another metabolic process that is essential for the function of the biosphere. Nitrogen (N_2) is the principal atmospheric gas and the element is part of the structure of proteins, nucleotides, chlorophyll, bacteriochlorophyll, and other vital molecules. Atmospheric nitrogen is useless for plants and cannot be assimilated by the majority of organisms. This is why nitrogen-fixing bacteria and archaea are so important. They enrich soils and aquatic habitats by converting N_2 into ammonia (NH_3). Free-living prokaryotes and symbiotic prokaryotes fix N_2 using the enzyme nitrogenase. This is a metabolically 'expensive' process because the nitrogen atoms in N_2 are connected via a strong triple bond that must be broken to reduce the molecule. Free-living cyanobacteria derive the necessary energy for nitrogen fixation from their photosynthetic activity, whereas non-photosynthetic bacteria that fix nitrogen within the root nodules of clover, soybean, and other legumes are supplied with energy by their plant hosts.

Chemoorganotrophs consume biological molecules produced by the solar- and mineral-powered organisms. Energy is tapped from these compounds by fermentation and respiratory reactions. Fermentation reactions provide cells with ATP (energy source) and NADH (reducing power) when oxygen is unavailable. One type of fermentation by bacteria and yeasts generates ATP and NADH by converting sugars to ethanol and CO_2; lactic acid is an alternative product of sugar fermentation. Respiratory reactions harvest more energy from sugars than fermentation by using electron transport chains like those that produce ATP and reducing power in autotrophs. Respiration can be aerobic or anaerobic. Oxygen is used in aerobic respiration to accept electrons at the end of the electron transport chain. In anaerobic respiration, organisms utilize alternative electron acceptors, including protons, carbon dioxide, and sulfate. The reduction of protons (H^+) produces hydrogen gas (H_2); carbon dioxide (CO_2) reduction by acetogenic bacteria yields acetate (CH_3COOH); sulfate (SO_4^{2-}) reduction generates hydrogen sulfide (H_2S).

Fermentation and respiration reverse the reactions used by autotrophs to form biological molecules.

Electron transport chains involved in respiration are situated in the cell membranes of bacteria and archaea. In eukaryotes, the electron transport chain used in respiration is located in a membrane that is folded inside the mitochondrion. This makes perfect sense in light of the symbiotic origin of the mitochondrion from a prokaryotic ancestor. Most eukaryotes can carry out aerobic respiration that results in the greatest energy yield from the oxidation of sugars. Modified forms of mitochondria called hydrogenosomes and mitosomes have evolved among anaerobic eukaryotes.

Sugars and other simple compounds consumed by chemoorganotrophs can be oxidized directly by fermentation or respiratory reactions. More complex molecules including polysaccharides, proteins, and lipids must be broken down into smaller molecules before they are absorbed into the cell. Digestive enzymes secreted by microorganisms release sugars, amino acids, fatty acids, and other smaller, soluble molecules from these macromolecules. These are absorbed by bacteria and fungi using transport proteins in their cell membranes and stored for future use, or oxidized by fermentation and respiratory reactions. Microorganisms use these digestive mechanisms to grow on surfaces or in fluids, and, in the case of filamentous fungi, when they penetrate solid materials.

Simple natural food sources for chemoorganotrophs include the carbohydrates in fruits. A fallen apple, for example, contains fructose, glucose, and sucrose. Fructose and glucose are monosaccharides which can be fermented directly to form alcohol and CO_2, or oxidized to release more energy through aerobic or anaerobic respiration. Bacteria and yeasts ferment these sugars under low oxygen conditions. Sucrose is a disaccharide made from fructose and glucose. To ferment sucrose, yeasts secrete invertase, which is an enzyme that cleaves the glycosidic bond in the sucrose molecule, releasing glucose and fructose. This is an example of the digestive processing that occurs before fermentation and respiration. Apples are also rich in cellulose, or fibre, which is a polymer of hundreds or thousands of D-glucose molecules linked in chains and grouped side by side into microfibrils that form plant cell walls. It is the most abundant organic polymer on Earth, present in standing vegetation, fallen trees, and plant debris in soils. Cellulose is degraded by multiple kinds of cellulase enzymes secreted by bacteria and fungi. Exocellulases remove glucose molecules from the ends of the cellulose chains and endocellulases chop away in the middle of the chains. Glucose residues released by cellulases fuel respiration by soil microorganisms that control the recycling of carbon in terrestrial ecosystems. As symbionts with animals, bacteria and fungi decompose cellulose in the guts of ruminants and termites.

Plant cell walls contain many other polysaccharides as well as the phenolic polymer lignin and proteins. All of these materials are digested by bacterial and fungal enzymes. Bacteria, archaea, and fungi grow in such a range of environmental conditions and produce such a broad array of secreted enzymes that very few naturally occurring and synthetic materials resist microbial digestion indefinitely. Digestive enzymes are used in a different way by amoeboid cells of eukaryotes. These cells feed by phagocytosis, surrounding bacteria and other microbes with their pseudopodia and absorbing them into food vacuoles in the cytoplasm. Enzymes secreted into the food vacuole digest the prey of the amoeba and the cell absorbs the sugars and other small molecules for aerobic or anaerobic respiration.

Microorganisms thrive in communities whose members complement one another by using different energy sources and creating local environmental conditions that offer mutual support. Microbial mats are dense assemblages of microorganisms that form in aquatic environments. Shallow salt water pools support a variety of microbial mats. Photosynthetic cyanobacteria absorb sunlight and produce oxygen at the surface of these communities. They also consume oxygen through aerobic respiration, depleting oxygen in deeper layers in the mat, which encourages the growth of anaerobic species. These include H_2S-oxidizing chemolithotrophs, photosynthetic green sulfur bacteria that are able to make use of the low light intensities inside the mat, chemoorganotrophs, and purple bacteria. Diatoms and other eukaryotic microorganisms coat the surface of the mat and penetrate the interior. Mobile mat inhabitants include cyanobacteria that migrate from the surface when the sunlight is most intense and can switch from oxygenic to non-oxygenic photosynthesis as the oxygen level plummets. Marine stromatolites are a form of microbial mat that creates a mineralized pillar as the lower layers of the microbial community are buried by accumulating sediment. The oldest fossilized cells are preserved in 3.5 billion-year-old stromatolites in Western Australia (Chapter 6).

Stromatolites are examples of biofilms. Less noticeable biofilms develop wherever bacteria, archaea, and eukaryotic microorganisms stick to surfaces and produce colonies. Biofilms cover the surfaces of plants, they coat rocks in aquatic habitats, and grow on our teeth; in the man-made environment, biofilms form a greasy sheen on boat hulls, shower curtains, and the interior of pipes. Channels run between the cells in the biofilm providing pathways for the transfer of gases and nutrients that nourish the community and for the release of its waste products. Environmental conditions vary within the biofilm, with nutrient levels modified by the residents and reduced oxygen availability in deeper locations favouring the growth of anaerobic prokaryotes. Bacteria signal to one another as they compete for space and nutrients and create an ordered colony structure in the biofilm. This molecular communication is a form of quorum sensing that microorganisms use to regulate gene expression according to population density.

Energized by the metabolic processes summarized in this chapter, bacteria, archaea, and eukaryotic microorganisms grow and reproduce, filling the biosphere with populations of dispersed cells, biofilms, and symbioses with other forms of life. Cell division in prokaryotes is an asexual reproductive process that increases population size. The time that elapses between each cycle of cell division is called the generation time. This varies greatly among bacteria and archaea. Our gut bacterium, *Escherichia coli*, divides every twenty minutes, offering an extreme contrast to methane-generating archaea in frozen Arctic lakes that double once per month. (Prokaryotes buried in seafloor sediments may show even less vigour, perhaps doubling on a timescale of years and earning the title 'slo-mo life'.)

Bacterial populations grow exponentially with the consequence that huge numbers of cells are produced when nutrients are plentiful. This is evident in the rapid progression of bacterial infections and the spoilage of food when a refrigerator stops working. Exponential growth is limited by the exhaustion of resources and the accumulation of toxic waste products from the burgeoning population of cells. The expansion of a population of cultured bacteria follows a predictable pattern beginning with a lag phase, when the initial population of the cells accumulates the materials needed to support rapid growth and division, the exponential phase, a stationary phase as growth slows, and, finally, death (Figure 13). The duration of these phases is manipulated in biotechnological applications to increase the yield of manufactured products. As nutrients are depleted, a few groups of bacteria produce spores inside their cells (endospores) or in chains, or gathered at the top of stalked fruit bodies. These reproductive strategies enable *Clostridium* and *Bacillus* species, streptomycetes (actinobacteria), and myxobacteria (proteobacteria) to survive unfavourable conditions.

Rod-shaped prokaryotes elongate until dividing by binary fission (Figure 14). Following chromosome replication, the division of the

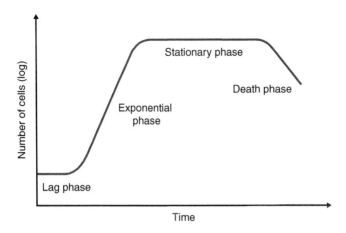

13. Graph illustrating the rise and fall of a bacterial population. A founding population absorbs nutrients and begins DNA replication in the lag phase. The number of cells increases through cell division and population size doubles at a constant time interval during the exponential phase, until critical nutrients become limiting and growth is slowed by the accumulation of waste compounds. The curve flattens in the ensuing stationary phase when cell division matches cell death. The population declines in the final death phase because food is exhausted

cell into equal halves is regulated by the formation of a ring of a protein called FtsZ that sits beneath the cell membrane. The constriction of this 'Z ring' coupled with the synthesis of new cell wall separates the daughter cells. FtsZ is produced in bacteria and archaea and is also involved in the division of mitochondria and chloroplasts in eukaryotic cells. Formation of the Z ring is inhibited until chromosome replication is completed and the two copies separate. Subsequent positioning of the Z ring is guided by proteins that function in locating the middle of the cell. FtsZ and associated proteins are described as the divisome complex. Components of the divisome may be good targets for novel antibiotics.

Cell division is linked to mitosis in eukaryotic microorganisms, but the mechanics of the process vary a lot according to cell

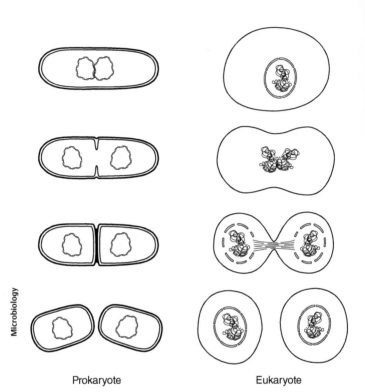

Prokaryote   Eukaryote

14. Cell division in (left) bacteria and (right) an amoeboid eukaryote

structure. The division of the amoeboid cells (and animal cells)
involves the constriction of a protein ring that forms at the equator
of the cell and pinches it into two halves (Figure 14). The position
of this ring is specified by the mitotic spindle that directs the
separation of chromosomes into the daughter cells. Distinctive
mechanisms of cell division occur in the fungi, including bud
formation and fission in yeasts, and the formation of cell
compartments along filamentous hyphae by the development of
septa (cross walls). Diatoms show one of the strangest mechanisms
of cell division in which the size of the average cell in a population

of these photosynthetic algae gets smaller with successive cycles of division. This is caused by the rigid cell wall structure of diatoms, which is organized as a lidded box. The wall is made from silica glass and mitosis is followed by the formation of the daughter cells inside the wall of the mother cell. Cell size is restored when diatoms undergo sexual reproduction and fashion both halves of the new shell.

Populations of yeasts expand in a similar fashion to bacteria, but the situation is more complex in filamentous fungi where hyphae within a feeding colony become internally divided by septum formation and cells branch to form a network of cells. There is no obvious way to relate the doubling time of a yeast colony to the growth of a fungal colony with cell compartments of differing length that contain varying numbers of nuclei. For practical reasons it is probably more useful to measure the increase in biomass of a filamentous fungus than to count its cell compartments.

Spore formation in response to nutrient limitation is also characteristic of the fungi. Some spores are adapted for dispersal from the parent colony to a location where the environment may be suitable for growth. Others are survival capsules that allow the individual to endure hostile environmental conditions until future conditions promote germination. Spores can be adapted for dispersal in air or in water. Spores vary greatly in shape and size and mycologists have used these microscopic characteristics as diagnostic features for identifying fungal species (Figure 15).

Fungi produce spores asexually and sexually. Chains or clusters of asexual spores, or conidia, develop on stalks that grow above the feeding colony. Nuclei in conidia are derived by mitosis which means that each spore is a clone of the parent colony that carries an exact copy of its genes. Different kinds of spores are produced following sexual recombination in fungi. Zygospores are sexual spores formed by zygomycete fungi. Mating between ascomycetes

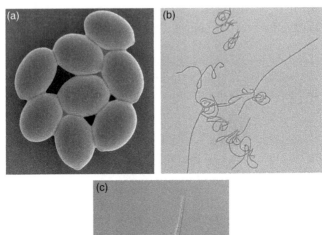

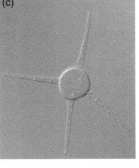

15. Selection of fungal spores illustrating some of the morphological variety that mycologists use to identify different species. (a) ascospores produced by a dung fungus; (b) filamentous spores of an insect parasite; (c) star-shaped spore of an aquatic fungus

produces ascospores and basidiomycetes produce basidiospores in mushrooms.

Airborne spores are not produced by most groups of eukaryotic microorganisms because their life cycles are aquatic. Dispersal in these organisms tends to occur via passive movement with the surrounding water. Resting spores, or cysts, serve as resistant structures that allow aquatic microbes to be dormant during winter months and to survive until the appropriate environmental conditions return. Much more is known about the life cycles of the fungi than other eukaryotic microorganisms because they can be cultured and have served as model experimental organisms for decades. The same is true for bacteria, which have been cultured, versus the majority of archaea that resist growth in the lab. The knowledge deficit is much greater than we have appreciated until recently, when genetic analysis of environmental samples began to highlight how many microorganisms have been overlooked because they fail to grow in the lab. The exploration of the metabolic function, cell biology, and behaviour of these cryptic life forms is one of the most exciting challenges in modern microbiology.

Chapter 3
Microbial genetics and molecular microbiology

The cell structures and metabolic processes described in Chapters 1 and 2 are specified by genes. Patterns of inheritance of these genes were unravelled using plants and animals as experimental subjects, but the study of genetics at the level of molecules has concentrated upon bacteria and fungi for more than sixty years. This work has been facilitated by the small genomes of microorganisms relative to multicellular species, the rapid generation times of cultured microbes, and the ease with which some fungi can be crossed to study the recombination of genes through sexual reproduction.

The genomes of bacteria, archaea, and eukaryotic microorganisms are encoded in double-stranded helices of DNA. Viral genomes are encoded in single-stranded and double-stranded DNA and RNA molecules. The expression of viral genes occurs within infected cells utilizing the molecular machinery of the host. Viral genetics is featured in Chapter 4. DNA and RNA are nucleic acids constructed from chains of nucleotides. Each nucleotide has a sugar molecule, a phosphate group that links the sugar molecule of one nucleotide to the next, and a nitrogen-containing base. (A chemical base is a compound that reacts with an acid to form a salt plus water.) The sequence of the bases in the nucleotide chain carries the genetic information. The first viral genomes, sequenced in the 1970s, comprised a few thousand nucleotides. The first genome of a cellular organism, the bacterium *Haemophilus influenzae*, was sequenced in 1995 and has 1.8 million base pairs. (We refer to base pairs because there are two complementary strings of nucleotides in double-stranded DNA.) The genome of *Methanocaldococcus jannaschii*, a thermophilic archaea that produces methane, is similar in size to *Haemophilus*. This was published in 1996 and was followed by the first eukaryote sequence, of the yeast *Saccharomyces cerevisiae*, which covered twelve million base pairs.

Advances in sequencing techniques and the development of automated sequencing methods have allowed scientists to sequence the genomes of 4,000 bacteria, 200 archaea, and 200 eukaryotes. Genome sizes vary a great deal within each category of microorganism and the largest prokaryote genomes overlap the smallest eukaryote genomes (Figure 16).

Highly automated methods of shotgun sequencing have been very effective for determining the complete DNA sequence of organisms. They involve cutting the genome into fragments, making copies of

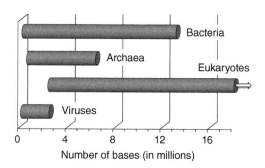

16. **Genome sizes for viruses and cellular organisms. The pandoravirus genome exceeds the size of some prokaryote genomes and overlaps the smallest eukaryotic genomes of parasitic fungi (microsporidians) that live inside other cells. The larger of the eukaryote genomes extend to billions of base pairs of DNA, way beyond the scale of this bar graph**

each fragment (cloning), sequencing the fragments, and assembling the genome by identifying overlapping sequences. The completed assembly shows the correct order of bases within the unfragmented chromosomes. Rapid developments in sequencing methods are producing faster 'reads' at lower cost. Analysis of the information in the genome begins with the search for sequences that may be translated into proteins. These sequences are called open reading frames (ORFs). ORFs are identified using a computer algorithm that finds the particular DNA sequences associated with the expression of genes.

Gene expression involves mechanisms that are common to all organisms as well as some processes that are specific to bacteria, archaea, and eukaryotes (Figure 17). The information encoded in the DNA molecule is transcribed into RNA molecules by the enzyme RNA polymerase (RNAP). RNA generated by transcription, called messenger RNA (mRNA), binds to ribosomes and is translated into proteins. The flow of genetic information from DNA

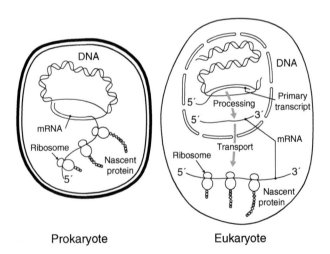

Prokaryote   Eukaryote

17. **Transcription and translation in (left) prokaryotes and (right) eukaryotes**

to RNA to protein was termed 'the central dogma' by Francis Crick in 1957 and the details of this intricate biochemical mechanism began to be unravelled through experiments on the gut bacterium *Escherichia coli* in the 1960s.

The bacterial polymerase is a large protein with five subunits. It binds to sites along DNA molecules called promoters that are located in advance (upstream) of genes. To allow transcription, the DNA double helix must open out to expose the nucleotide bases that specify the order of amino acids that are assembled into proteins. RNAP does this by advancing along the gene unwinding the double helix to create a 'transcription bubble' that rewinds in its wake. The polymerase uses one of the open strands of the DNA as a template for the synthesis of an RNA molecule whose nucleotide bases complement the nucleotide bases on the DNA. DNA is a polymer of deoxyribonucleotides that contain four kinds of nitrogen-containing base: adenine, guanine, thymine, and cytosine (A, G, T, C). RNA is a polymer of ribonucleotides whose bases are adenine, guanine, uracil (rather than thymine), and cytosine (A, G, U, C). A and G are types of base called purines; T, C, and U are pyrimidines. In the DNA molecule, As are linked to Ts in complementary strands by hydrogen bonds, and Gs are bonded to Cs. RNAP transcribes the DNA sequence into a single strand of RNA by matching As to Us, Ts to As, Gs to Cs, and Cs to Gs and producing a sequence of ribonucleotides linked by covalent bonds. RNAP is released when the enzyme encounters a sequence called the transcription terminator.

Most genes specify proteins, but ribosomal RNA (rRNA) and transfer RNA (tRNA) are generated by transcription and are not translated. Three types of rRNA combine with multiple proteins to form the structure of bacterial ribosomes. Sixty kinds of tRNA function in the translation mechanism by adding amino acids to growing polypeptide chains that form proteins. Genes encoding different kinds of rRNA and tRNA are transcribed as single units in bacteria and are cut into separate RNAs after transcription.

These genes (also called cistrons) are described as polycistronic. A group of these cotranscribed genes controlled by a single promoter constitutes an operon. Most of the messenger RNA produced by transcription in archaea is also polycistronic.

Translation of mRNA into proteins occurs when mRNA binds to ribosomes. The mRNA sequence specifies the sequential addition of amino acids to polypeptide chains that form proteins. Different amino acids are encoded as triplets of nucleotide bases in the mRNA sequence called codons. In addition to encoding amino acids, triplets specify positions where translation begins (start codons) and where translation is terminated (stop codons). Each of twenty-two amino acids is encoded by one or more codons according to the genetic code. The code is almost universal, but there are some variations with certain organisms using stop codons to encode amino acids. Codons are complemented by anticodons that are integrated into the sequence of different kinds of cloverleaf-shaped tRNA molecules that carry amino acids to the ribosome. The codon UAU on the mRNA sequence is complemented by a tRNA bearing the anticodon AUA that carries the amino acid tyrosine. The match between the anticodon of the tRNA and the amino acid that it carries ensures that the polypeptide chains are assembled according to the order of codons in the mRNA sequence.

Transcription is regulated by proteins that bind to DNA sequences. The DNA of eukaryotes and many of the archaea is spooled around proteins called histones allowing the DNA to be compacted into a relatively small space. DNA combines with complexes of different kinds of histone to form nucleosomes that look like beads on a string and this structure condenses upon itself to create a denser fibre. Further packaging of this fibre produces the distinctive form of multiple chromosomes in eukaryotic cells when they are undergoing cell division. In this coiled state, the RNA polymerase enzyme is unable to transcribe the DNA sequence into mRNA. Gene expression is dependent upon

Microbiology

uncoiling and this is catalysed by enzymes that reverse the coiling process. The absence of histones in bacteria simplifies these processes.

Most of the proteins involved in gene regulation bind to specific DNA sequences and affect the transcription of single genes or an operon of cotranscribed genes in the same location on the chromosome. Some genes are transcribed all the time, providing a constant supply of mRNA which is translated into protein. This is called constitutive expression. Other genes are turned on and off in response to the requirements of the cell and changes in environmental conditions.

The classic illustration of gene regulation in bacteria is the mechanism that controls the levels of three enzymes (lacZ, lacY, lacA) involved in lactose utilization. Lactose is a disaccharide sugar found in milk. The genes encoding the enzymes are next to one another on the bacterial chromosome and form an operon controlled by two upstream sequences called the promoter and operator. When RNAP binds to the promoter the three *lac* genes are cotranscribed to form a single polycistronic mRNA molecule. In the absence of lactose, the bacterial cell conserves energy by limiting transcription of the operon. This is achieved by the binding of a repressor protein to the operator sequence that follows the promoter, which interferes with RNA polymerase binding. When lactose is available, a breakdown product from the sugar binds to the repressor and prevents it from binding to the operator. This allows RNAP unimpeded access to transcribe the operon. Regulation of the *lac* operon was investigated in the 1960s using mutants of the bacterium *Escherichia coli* expressing disabilities in lactose metabolism. Transcription of many other genes is activated rather than repressed. These positive control mechanisms involve activator proteins that stimulate RNAP binding to DNA.

Gene expression is also controlled by more complex regulatory mechanisms involving sensory proteins (SKs) that detect environmental signals, and response regulator proteins (RRs) that govern transcription. These proteins constitute a two-component regulatory system. The sensory protein is a type of enzyme called a histidine kinase that is situated in the cell membrane. When the kinase detects the appropriate signal it transfers a phosphate group to a histidine residue (amino acid) within the core of its structure and then transfers this phosphate group to the response regulator protein in the cytoplasm. This reaction causes a change in the shape of the response regulator protein that stimulates or represses the transcription of target genes.

Some bacteria have more than a hundred of these signal transduction mechanisms, equipping them to respond to changes in salinity, acidity, temperature, nutrient availability, and other environmental challenges. Bacterial movement toward chemical attractants and away from repellents is controlled by a sophisticated two-component regulatory system that affects the flagellar motor. The detection of these chemicals in the fluid surrounding the cell activates a sensor kinase. This enzyme triggers a response regulator protein that controls the direction of rotation of the flagellar motor. Anticlockwise rotation of the motor drives the cell in a straight line; clockwise rotation causes the cell to tumble and reorient before swimming away in another direction. This mechanism, referred to as chemotaxis, is a relatively simple illustration of the complex relationship between individual proteins encoded in the microbial genome and the behaviour of the cell. Two-component regulatory mechanisms involving histidine kinases are uncommon in eukaryotes but they have been identified in fungi and other eukaryotic microbes.

The heat shock response offers another example of the genetic control of cell activity. Heat shock proteins (HSPs) are produced by all organisms when they are stressed by heat and other factors that compromise their biological activities. HSPs act as chaperones, maintaining the correct folding (conformation) of proteins and preventing protein aggregation. The transcription of genes encoding HSPs in bacteria and archaea is regulated by small proteins (sigma factors) whose levels respond directly to temperature. HSPs allow prokaryotes to survive temperature spikes and chemical treatments that would otherwise inactivate proteins. Homologous HSPs in humans are important in maintaining cardiovascular health and play essential roles in the immune system.

The genome of bacteria and archaea is organized as a single circular molecule of DNA. Replication of this chromosome begins at a single location, called the origin of replication, and proceeds in both directions around the loop. Copying of the DNA molecule is catalysed by DNA polymerases that build new strands of nucleotides on both strands of the existing DNA double helix to create a pair of complete chromosomes. Replication in *Escherichia coli* is completed in about forty minutes with the polymerases adding 1,000 nucleotides per second. Other enzymes involved in replication unwind the double helix in advance of the polymerases (helicase) and seal the opening in the new strand at the completion of synthesis (DNA ligase). Errors in the replication process are a source of mutations and are often lethal for the cell. The accuracy of replication is maintained by an active proofreading system in which the polymerases respond to distortions in the DNA helix caused by the incorporation of an incorrect nucleotide by replacing the errant nucleotide with one carrying the correct base.

Replication in archaea proceeds from two or more origins of replication on a single chromosome. This is one of the many distinctions between the molecular genetics of bacteria and archaea. Features of archaea that are more similar to eukaryotes than bacteria include the homology between the single type of RNA polymerase in archaea and one of the RNA polymerase enzymes used by eukaryotes, the promoter sequences of archaea and eukaryotes, and their translational machinery.

There are, however, many differences between the genetic systems of archaea and eukaryotes, beginning with the isolation of multiple linear chromosomes within the nucleus of eukaryotes. The nucleus is a game changer for the cell because it separates the mechanisms of messenger RNA synthesis (transcription) and protein synthesis (translation). Messenger RNA must move through pores in the nuclear envelope for translation by ribosomes in the cytoplasm (Figure 17). This provides the eukaryote cell with greater control over protein synthesis. Eukaryote genes are organized differently from those of prokaryotes. Operons are uncommon in eukaryotes. The presence of non-coding regions of DNA (introns) between coding regions (exons) of eukaryote genes results in the formation of mRNA molecules containing the sequences of introns and exons. The introns are removed from the nascent form of the mRNA, the exons are spliced together, and the processed mRNA moves from the nucleus into the cytoplasm where it is translated into protein. The job of cutting and splicing RNA is done by a complex of RNA and proteins called the spliceosome. Introns are very rare in bacteria and archaea. Eukaryotic mRNA is also modified by the addition of a 'cap' at the end of the molecule where translation begins (identified as the 5'-end) and synthesis of a 'tail' of 100–200 adenylate (A) residues at the other end (3' end). The addition of the cap and poly(A) tail is essential for mRNA export from the nucleus, for initiating translation, and other purposes. Short poly(A) tails are also added to mRNA in prokaryotes and in mitochondria and chloroplasts of eukaryotes.

Biologists refer to the physical manifestation of genes as phenotype and the complete suite of genes as the genotype. The mechanistic links between the genotype and phenotype are understood much better in single-celled microorganisms than they are in multicellular species. Gene expression can be studied using DNA microarrays of gene segments fixed to a solid chip made from glass, plastic, or silicon. Millions of copies of each DNA sequence are dotted onto the chip in a specific two-dimensional pattern and the investigator can study which of the genes are being transcribed under a particular set of conditions by adding a labelled solution of the mRNA from a population of cells. The mRNA is labelled with a fluorescent dye, and after this sample has been allowed to react with the microarray, mRNAs bound to the segments are identified from the positions of their complementary genes on the chip. This snapshot of transcription is one of the sources of information used in the study of the entire complement of RNA. This is called the transcriptome. Mass spectrometry is used to study the proteins produced by translation of the mRNA (the proteome) as well as the metabolites generated by the activity of the proteome (the metabolome). These powerful techniques yield large and complex data sets whose analysis requires experts in computer science specializing in the interdisciplinary field of bioinformatics. Transcriptomic, proteomic, and metabolomic studies go hand in hand with genomic analysis, affording information on the utilization of the genome as the organism undergoes developmental changes and encounters different environmental conditions.

More than 90 per cent of the genomes of bacteria and archaea code for proteins. Prokaryotes that live inside the cells of eukaryotes as damaging parasites and supportive 'endosymbionts' have the smallest genomes of cellular organisms. Myxobacteria have some of the largest prokaryote genomes. The number of protein-encoding genes in prokaryote genomes varies from 169 in the insect endosymbiont *Hodgkinia cicadicola* to 11,599 in a myxobacterium called *Sorangium cellulosum* that lives in soil. Massive gene loss is characteristic of endosymbiotic prokaryotes. This is reflected in the tiny genomes of the mitochondria and chloroplasts of eukaryotes that evolved from free-living bacteria. Experiments involving the disruption of genes within the already tiny genome of *Mycobacterium genitalium* predict that a minimum of 250–300 genes are essential for survival of this bacterium.

Eukaryote genomes contain much higher proportions of non-coding sequences, making it impossible to estimate the number of functional genes from the number of base pairs of DNA. The yeast genome is relatively compact for a eukaryote, with one-third of its DNA occupied by non-coding sequences. It is organized into sixteen chromosomes and encodes 6,000 functional genes. Thirty per cent of the yeast genes are homologous to human genes, meaning that genes with similar sequences are found in both organisms and seem to share descent from the same ancient ancestor. The genome of the social amoeba *Dictyostelium discoideum* is three times larger than yeast, encodes twice the number of proteins, and almost 40 per cent of its DNA is non-coding. Other eukaryotic microorganisms have more cluttered genomes. The massive genome of the amoebozoan *Amoeba proteus* is packaged into 500–1,000 chromosomes and the genome of its relative, *Polychaos dubium*, is estimated to be 200 times bigger than the human genome. It is assumed that much of the genome of *Polychaos* is non-coding. The alterative hypothesis is that it takes a much greater genetic endowment to be an amoeba than a human.

Some of the DNA that does not encode proteins is transcribed into functional forms of RNA involved in protein synthesis (ribosomal RNA and transfer RNA) and RNA molecules that regulate the expression of protein-encoding genes. Other non-coding sequences called introns are inserted within functional genes. DNA that does not appear to encode proteins or functional RNA molecules has been referred to as 'junk DNA'. Some of these sequences may be redundant, coding for genes (pseudogenes) inactivated during the evolutionary history of an organism's genome or introduced by viruses and performing no current function. This is an active area of research. The analysis of whole genomes is one of the primary missions of modern microbiology. Genomic research provides information on disease-causing mechanisms, ecological activities of microbes, the biotechnological potential of different species, and evolutionary history.

All genetic material is vulnerable to mutation. Natural mutations in microorganisms play a primary role in evolution. Point mutations within coding sequences affect single base pairs. These base-pair substitutions cause a range of effects by interfering with transcription or translation and can produce non-functional proteins. If the mutation produces a triplet that specifies the same amino acid as the original sequence, the mutation will have no effect on protein synthesis. These mutations are silent. The insertion or deletion of base pairs into the coding regions of the genome, or the deletion of sequences, tend to be very damaging. This is because their effects on gene expression cannot be reversed by a simple back mutation that restores the original sequence of base pairs. Deliberate mutation of microorganisms using chemical mutagens, ultraviolet light, and ionizing radiation has been a widespread research practice for decades. These methods damage the genome in a random fashion and are used to create 'libraries' of mutant genes. Directed mutation of specific genes, known as site-directed mutagenesis, has become the more common method in modern molecular genetics and genetic engineering.

Natural mechanisms of genetic modification have evolved among bacteria, archaea, and eukaryotic microbes. Transformation occurs when DNA fragments released from a donor cell are incorporated into a recipient cell. Integration of the DNA into the chromosome of the recipient cell is mediated by the protein RecA. Transformation does not happen frequently in nature, but it can be stimulated under experimental conditions. Transduction is a second mechanism of genetic modification in which a virus transfers DNA from one cell to another. This is thought to be more important as a natural mechanism. Mating of bacteria allows for the transfer of DNA between cells through a tubular structure called a sex pilus. This genetic transaction involving physical contact between cells is called conjugation. The pilus is formed from multiple protein subunits. The mating cells are pulled together by the disassembly of the pilus subunits and form a pore at the contact site through which the DNA is transferred.

Plasmids are often transferred during conjugation. Plasmids are small circular DNA molecules that encode accessory genes that are not found on the chromosome. A single cell can contain more than a hundred copies of a single type of plasmid. Genes on resistance plasmids encode proteins that protect the cell against antibiotics. Other kinds of plasmid affect the virulence of a bacterial strain by enhancing its ability to colonize and damage a particular host.

Sequences of DNA that can move from one location within a genome to another are called transposable elements or transposons. They are found in all organisms and are important features in genome evolution. In prokaryotes, transposons can move from plasmids integrating antibiotic resistance into the chromosome. If the transposon is not copied before transposition (cut and paste), the sequence is lost from its original location; if the transposon is copied before transfer (copy and paste) the original sequence remains in the donor site. Most transposons do not cause any damage to the cell, but when transposons land within functional genes they are likely to disable them.

Reproduction in prokaryotes is always asexual. Conjugation and other mechanisms exchange genes between cells, but bacteria and archaea do not engage in the wholesale pairing and recombination of genomes that occurs in sexual reproduction in eukaryotes. Sexual reproduction is common among eukaryotic microbes. Cells of the yeast *Saccharomyces cerevisiae* have a single set of chromosomes. The single nucleus of each cell divides by mitosis and one of the daughter nuclei is shifted into the bud (Figure 18). Each cell may produce a series of buds that leave scars on the surface of the mother cell upon separation. There are two mating types in *Saccharomyces*, designated alpha and 'a'. Cells of opposite mating type fuse to produce a single cell with two sets of chromosomes (diploid cell) and meiosis produces four spores with a single set of chromosomes. Meiosis is the type of cell division that produces sperm and egg cells in animals. *Saccharomyces* is an ascomycete fungus and its sexual spores are called ascospores.

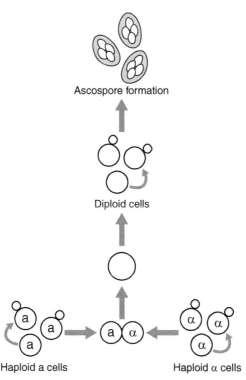

Ascospore formation

Diploid cells

Haploid a cells   Haploid α cells

18. **Sexual reproduction in baker's yeast, *Saccharomyces cerevisiae***

Ascospore germination releases a fresh yeast cell that grows and divides by bud formation. More complex life cycles are common among filamentous fungi. Zygomycete fungi produce colonies of filamentous hyphae and produce sexual spores called zygospores by conjugation between colonies of different mating types (Figure 19). There are a number of steps in this process that are controlled by volatile pheromones. Compatible colonies of mushroom-forming basidiomycetes fuse to produce filamentous hyphae with cell compartments containing a pair of nuclei of each of the mating types. These nuclei remain separate until the colony

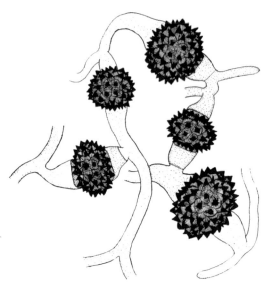

19. Sexual reproduction in a filamentous zygomycete fungus,
Cokeromyces recurvatus

forms a mushroom and fusion occurs on cells on the surface of the
gills. Meiosis follows fusion and the four daughter nuclei are
packaged into spores that are propelled from the gills.

Mechanisms of sexual reproduction have been studied in
eukaryotic microorganisms belonging to all of the supergroups
described in Chapter 1. The social amoeba *Dictyostelium* has a
sexual cycle in which cells of opposite mating type fuse and then
cannibalize other cells in the population to support the
development of a cyst from which the next generation of
recombinant cells is released. Absence of evidence for
recombination in other species is not evidence of absence: sexual
reproduction may occur in the ostensibly celibate amoebozoans
under conditions in nature that have not been replicated in
the lab. Genomic analysis can reveal these cryptic sexual cycles.
The genome of the amoebozoan *Entamoeba histolytica*, which is

an intestinal parasite, includes genes for meiosis suggesting that it has a mechanism of cell fusion.

In addition to genetic research on single species, microbiologists have developed methods for exploring the diversity of microorganisms in different environments. This metagenomic research has revealed an unexpected profusion of organisms in aquatic ecosystems and soil as well as the human digestive system. This work proceeds without looking at samples with a microscope or culturing microbes in the lab. Using methods of shotgun sequencing, metagenomic research provides sequence information derived from the totality of the DNA within a sample. These experiments identify huge numbers of genes that show little or no similarity to known genes and reveal organisms that are new to science.

Chapter 4
Viruses

Most viruses are dwarfed by the cells that they infect and viral populations represent a small fraction of the total biomass in any ecosystem. Yet viruses are the most numerous carriers of genetic information. In the deep ocean, for example, viruses outnumber cellular forms of life by a ratio of 100 to 1. Pathogenic viruses are an omnipresent and unpleasant part of our lives, causing infectious illnesses ranging from the global distractions of the common cold to the horror of Ebola outbreaks in Africa. Routine vaccinations to prevent a range of viral illnesses are one of the blessings of modern medicine and powerful therapies have been developed to treat the symptoms of many kinds of chronic viral infections including hepatitis B and C, oral and genital herpes, and the human immunodeficiency virus (HIV). Nevertheless, millions of people die from interacting with viruses every year.

Beyond the effects of viruses on human health, the lives of all organisms and the cycling of nutrients through the biosphere depend upon the activities of viruses. Viruses control populations of bacteria, archaea, and eukaryotes and this destructive power liberates massive quantities of nutrients in aquatic and terrestrial ecosystems. The diversity of viruses revealed by metagenomic research is quite staggering and recent discoveries have altered our ideas about the limits of viral complexity. The ubiquity and prevalence of viral genes in the genomes of cellular organisms is a measure of the influence of viruses on the evolution of prokaryotes and eukaryotes and we are only beginning to grapple with the origin of viruses themselves.

The science of virology is younger than the field of bacteriology. The existence of viruses was predicted after scientists had already made tremendous progress in linking particular bacteria to specific infectious diseases. Applying Koch's Postulates (Chapter 1), bacteria were isolated from infected laboratory animals, grown in culture, and injected into healthy animals. Subsequent development of the disease demonstrated that the cultured bacteria were the infectious agents. This procedure failed for smallpox, rabies, and many other diseases, indicating that something else was responsible for infection. The reason that the experiments failed was that viruses cannot grow in pure culture: they are reliant upon access to living cells in which they reproduce.

Even though the physical nature of viruses was unknown until the 20th century, skin inoculation with smallpox scabs was practised thousands of years ago in India and China as a preventive against the disease. Edward Jenner developed the safer use of scratchings from cowpox lesions to vaccinate against smallpox in the 1790s, and a century later Louis Pasteur introduced a vaccine against rabies prepared from infected rabbit brains. The breakthrough in understanding viruses came at the end of the 19th century from experiments showing that some infectious agents were not removed from fluids strained with fine filters to exclude bacteria. The earliest of these investigations were conducted in Holland on the tobacco mosaic virus (TMV). This virus ruins tobacco crops by causing leaf blotching and wrinkling. The first images of viral particles using the electron microscope and experiments on the chemical composition of viruses were published in the 1930s.

The structure of a virus is much simpler than the structure of a prokaryote cell. The viral genome is encoded in one or more DNA or RNA molecules and these nucleic acids are surrounded by a protein coat called the capsid (see Figure 8 in Chapter 1). The capsid is composed of a number of protein subunits or capsomeres. In addition to the nucleic acid and protein components, a lipid envelope that resembles a cell membrane surrounds the capsid of many viruses. Other components can include enzymes that are packaged inside the capsid. The whole structure of the virus particle is referred to as a virion.

Capsomere subunits are assembled into icosahedral and helical arrangements to create a range of capsid shapes and sizes. Variation on the icosahedral form arises from the way in which the twenty triangular faces of the structure are produced. For example, if three protein subunits form each face, the capsid will comprise a total of sixty subunits. Most virus capsids are assembled from more than sixty subunits. A remarkable feature of capsid assembly is that it doesn't require any input of metabolic energy by the host cell. Once the components of the particle are synthesized, these building blocks bind to one another in a highly ordered fashion through a mechanism of self-assembly. The 'growth' of the virus is like the physical process of crystallization.

Simple capsids are built from multiple copies of the same protein; more complex structures assembled from copies of several different proteins are more common. The rod-shaped tobacco mosaic virus has a helical capsid structure composed of 2,130 subunits of a single protein. Self-assembly of TMV particles occurs in a simple mixture of protein subunits and the RNA genome of this virus. Seventeen of the subunits bind to one another forming a flat two-layered disc with a hole in the middle. A hairpin loop toward one end of the RNA molecule slips into the hole in the disc and binds to the gap between the two layers of protein in the disc, causing the disc to dislocate into the shape of a split washer (or lock washer). This structure is the foundation for the helical conformation of the mature TMV particle, whose assembly proceeds with subunits stacking around the RNA molecule until the entire length of the genome is enclosed.

Picornaviruses, including rhinoviruses and the poliovirus, form more complex icosahedral capsids using sixty copies each of four different proteins. These are preassembled into groups of three proteins that form larger pentamers (with a total of fifteen protein subunits) before incorporation in the full capsid. Rather than protein accumulation around the genome, which happens in TMV replication, the RNA genome of picornaviruses is added to the particle after assembly of the capsid. Adenoviruses are more complex again, building their icosahedral capsids from seven different proteins including one subunit that forms extended fibres that project from the corners of the particle.

In addition to the proteins that form the capsid surrounding the genome, the interior of the mature particle may incorporate viral enzymes. These enzymes perform a number of functions during the next phase of infection and replication. Viruses equipped with envelopes acquire these lipid bilayers from the membranes of their host cells during the exit process. Viral envelopes can contain proteins and glycoproteins (proteins festooned with polysaccharide chains) encoded by the viral genome. These molecules bind to receptor proteins on the host membrane and facilitate viral uptake. In addition to their involvement in the entry process, glycoproteins in the envelope enable viruses to evade the immune defences of the host. One way in which this is done is called glycan shielding. This refers to the way in which the chains of sugar molecules of glycoproteins interfere with host antibodies, preventing them from recognizing proteins on the envelope surface.

The importance of glycoproteins in virus entry and evasion of the host immune system is illustrated by HIV. The envelope of HIV is ornamented with molecular spikes assembled from a pair of glycoproteins called gp41, which anchors the structure in the envelope, and gp120, which is exposed at the surface. Fusion of the HIV envelope with the membrane of the host cell is dependent upon the binding of gp120 to a receptor protein on the surface of the host cell called CD4. Applying a lock and key analogy to this molecular interaction, we can view gp120 as the key that opens the lock, represented by CD4, to enter the host cell. In addition to its role in viral entry, gp120 obscures potential binding sites on the viral surface from antibodies, allowing HIV to escape immune recognition. This glycoprotein is essential for replication of the virus and is an important target for vaccines against HIV. Unfortunately, the diversity of glycoprotein structure introduced by the continuous evolution of the virus creates a moving target for researchers. Glycoproteins of other viruses interrupt a variety of defensive processes utilized by host immune systems.

The requirement for replication within living cells of a host organism is a unifying characteristic of viruses. Cells of prokaryotes and eukaryotes multiply by making a complete copy of their genetic material and dividing by the mechanisms of binary fission and mitosis. This is not an option for a virus. Limited by their relatively simple structure, viruses reproduce by hijacking the molecular machinery of nucleic acid and protein synthesis inside the living cells of every kind of organism. There are five steps in the mechanism of viral replication: (1) attachment, (2) penetration, (3) synthesis, (4) assembly, and (5) release (Figure 20). The time taken to complete all of these steps varies from twenty minutes in viruses that infect bacterial cells to forty hours for the slowest of the animal viruses.

Attachment depends upon a specific interaction between proteins, glycoproteins, or lipids on the surface of the virus and receptor molecules on the surface of the host cell. The match between receptor and virus must be perfect, as we have seen with HIV, and this is one of the reasons that particular viruses only attack the cells of specific hosts. After attachment of a virus, its genome, or the whole viral particle, enters the host cell. This is the penetration phase. Viruses called bacteriophages, which infect bacteria, stay on the outside of the host cell and inject their DNA into their hosts. The complex shape of some of the bacteriophages

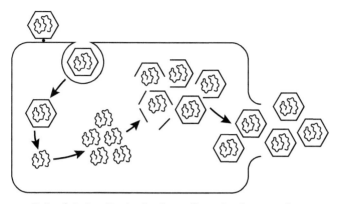

20. Cycle of viral replication in a host cell running from attachment to the cell surface, penetration by endocytosis, exposure of the viral genome, replication (synthesis) of the genome and capsid components, assembly of the next generation of virus particles, and release by lysis of the host cell

recalls the Apollo Lunar Module (Figure 21). The penetration process is completely different for many animal viruses because the entire particle—protein capsid and the viral genome—is absorbed by the host cell through the mechanism of endocytosis (as illustrated in Figure 20). Once inside the host cell, the protein coat of these animal viruses is removed, exposing the genome. In the synthesis phase, proteins encoded by the viral genome are manufactured by the host cell.

Two types of virus protein are synthesized by the host. These are proteins needed for copying the viral genome, and structural proteins that form the virus particles. Viral infections disrupt the normal functioning of the host cells and inhibit the usual production of host proteins in favour of generating the components needed for the new generation of viruses. In many infections the viral genome encodes proteins that sustain the parasitic relationship by ensuring that the host cell continues to operate as long as possible while it is engaged in virus production.

Assembly of the protein capsids and packaging of the copies of the DNA or RNA genomes into the particles occurs inside the host cells. Once the assembly phase is completed, release of hundreds or thousands of viruses can occur by bursting of the host cell or by a more gentle form of escape via exocytosis. Tailed bacteriophages encode hydrolytic enzymes called endolysins that dissolve the peptidoglycan wall of their bacterial hosts.

Viruses that infect multicellular organisms use a variety of mechanisms to enter the tissues of their hosts before they penetrate susceptible cells. These include introduction through a wound in the skin of an animal or an abrasion on the surface of a plant. Insects and ticks can act as vectors for the transmission of animal and plant viruses from one host to another. Yellow fever is spread by mosquitoes and causes tens of thousands of deaths every year. Most cases of yellow fever occur in Africa. Other mosquito-borne viral diseases include dengue fever, which is a widespread tropical disease, and West Nile fever, first identified in Uganda, which was introduced into the United States in 1999. Yellow fever, dengue fever, and West Nile are all caused by RNA viruses classified as flaviviruses. Many other human viruses infect animal tissues directly following inhalation or via contact with infected body fluids through sexual activity or following injection with contaminated needles. Aphids are important vectors for the transmission of plant viruses.

Because they cannot be cultured on their own, virologists raise viruses in cultures of microorganisms, cultures of animal cells, and in whole animals and plants. To identify and to count bacteriophages, samples containing these viruses are mixed with pure cultures of bacteria and melted agar and poured on top of a solid agar surface. The melted agar solidifies after pouring and the bacteria divide and form an even lawn of cells within this top layer during an overnight incubation. Viruses that infect the cultured cells become visible as clear areas, or plaques, of dead bacteria. Each of the plaques can be initiated by a single virion as it multiplies through successive cycles of replication. The number of plaques, or plaque-forming units, and the dilution factor are used to calculate the number of viruses in the original sample. Monolayers of cultured animal cells are used widely to study animal viruses, and mice and other laboratory animals are used when cell cultures are ineffective. Cultures of plant tissues as well as whole plants are used to study plant viruses, with none of the attendant ethical issues associated with research on animal virology.

The organization of viruses into different groups is based upon the type of genome and its mechanism of replication. This seemingly esoteric method considers the fundamental genetic structure of the virus rather than its shape and size, its host organism, or the type of disease that it causes. There are seven groups identified by Roman numerals (Table 1). The genome of Group I viruses is encoded in a double-stranded molecule of DNA (dsDNA) housed inside the capsid. Herpesviruses and poxviruses are in this group. The DNA of these viruses is transcribed into mRNA in the same way that the DNA of the host cell is expressed (Chapter 3). Group II viruses contain single-stranded DNA (ssDNA) and the missing complementary strand of nucleotides must be synthesized after cell infection because the host's RNA polymerase enzyme can only generate RNA by unwinding and transcribing a DNA double helix. Group II viruses include parvoviruses that cause deadly diseases in cats, dogs, and other mammals, but do not cause serious illnesses in humans.

The genes of the next four groups of viruses are encoded in RNA rather than DNA. Reoviruses and other Group III viruses contain double-stranded RNA (dsRNA). Reoviruses include rotavirus that causes a common form of gastroenteritis in children, and viruses that cause infections in the respiratory tract. Their dsRNA genomes are peculiar molecules: cells do not produce any double-stranded RNA. Expression of reoviral genes is, however, a straightforward process. When DNA is transcribed into mRNA, one

Table 1. Classification of viruses according to the structure of their genomes

Group	Genome	Examples
I	double-stranded DNA (dsDNA)	herpesviruses, poxviruses
II	single-stranded DNA (ssDNA)	parvoviruses
III	double-stranded RNA (dsRNA)	reoviruses
IV	single-stranded, positive strand RNA (+ssRNA)	coronaviruses, picornavruses
V	single-stranded, negative strand RNA (−ssRNA)	rabies virus, filoviruses, paramyxoviruses
VI	single-stranded, positive strand RNA-RT (ssRNA-RT)	retroviruses
VII	double-stranded DNA-RT (dsDNA-RT)	hepatitis B virus

RT, reverse transcriptase

of the strands of the DNA double helix serves as the template for the formation of the RNA molecule. The DNA strand that is read into mRNA is called the antisense or negative strand and the transcribed mRNA is considered a positive strand. The double-stranded RNA of a reovirus has a positive and negative strand and its negative strand is transcribed into mRNA in the host cell.

Group IV viruses have a single strand of positive strand RNA (+ssRNA), which is used directly as the mRNA that is translated into viral protein. Coronaviruses, including the severe acute respiratory syndrome (SARS) coronavirus, and picornaviruses that cause polio and the common cold are examples of Group IV viruses. The single-stranded RNA of Group V viruses is a negative strand RNA genome (−ssRNA) that is transcribed into mRNA. Group V viruses include the rabies virus, filoviruses that cause Ebola and Marburg haemorrhagic fevers, and paramyxoviruses that cause measles and mumps.

Viruses classified in Class VI and Class VII have the most complex mechanisms of gene expression. Retroviruses, including HIV, are Class VI viruses containing single-stranded positive RNA molecules. Rather than direct use of the Class VI genome as a form of mRNA, the retroviral RNA is converted into a double-stranded DNA molecule following cell invasion. These Class VI viruses are identified as ssRNA-RT, where the RT refers to a process called reverse transcription. Normal transcription makes an RNA copy of a DNA sequence; reverse transcription makes a DNA copy from an RNA template. The enzyme that carries out this synthesis, called reverse transcriptase, is carried into the host cell within the virus particle. The DNA copy of the viral genome is incorporated into the genome of the host cell by an integrase enzyme encoded by a retroviral gene. Once this happens, the retrovirus is replicated along with the rest of the host DNA and carried as a persistent viral infection. Following integration, the retrovirus can remain as a latent or lysogenic infection, in which its genes are not transcribed, or become productive, resulting in the generation of new viruses. If the DNA of retroviral origin is transmitted to the egg or sperm cells of an animal, it can be inherited by the next generation. This mechanism is responsible for the prevalence of endogenous retroviruses in the human genome.

Viruses assigned to the final class, Class VII, have a double-stranded DNA genome (dsDNA-RT) that is used as a template for the synthesis of RNA that is reverse transcribed into the replicated DNA genome. Hepatitis B is a disease caused by one of these bizarre viruses whose genes are transmitted from DNA to RNA to DNA during the replication cycle. It is an example of an oncovirus, which is a virus that causes cancer. The link between persistent infection with hepatitis B virus and cancer is clear from epidemiological studies showing that countries where the infection is most prevalent also show a high incidence of liver cancer. When the virus is active it destroys liver cells and causes inflammation. The induction of cancer is indirect, with the involvement of a specific viral gene (gene X) affecting the expression of genes in the host cells that control cell division. Liver cancer is also associated with hepatitis C, which is a +ssRNA virus (Class IV). Other oncoviruses include the human T lymphotropic virus (HTLV-1), a retrovirus that infects T cells of the immune system, the human papilloma virus that causes cervical cancer, and Kaposi's sarcoma-associated herpesvirus whose effects are commonly associated with the damage to the immune system caused by HIV infection (Chapter 5). Oncoviruses have been implicated in 10–20 per cent of all cancer cases.

The smallest viruses include ssDNA parvoviruses and microviruses, +ssRNA picornaviruses, and, tiniest of all, the ssDNA porcine circovirus which has a diameter of only 15–20 nanometres (nm), or one-sixtieth the size of the average bacterial cell. This circovirus causes a common wasting disease in pigs. Its genome encodes two genes: *rep* that is translated into a pair of proteins, called replicases, which copy the viral genome in the host cell, and *cap* that encodes a single kind of capsid protein. Sixty copies of the capsid protein are assembled into the icosahedral form of the circovirus. Viruses do not get any simpler than this.

The +ssRNA of the coronavirus that causes SARS is packaged in a 90 nm diameter capsid surrounded with a lipid envelope decorated with glycoprotein spikes. Its genome is among the largest among the RNA viruses and encodes fourteen proteins. Herpes simplex viruses that cause cold sores (HSV-1) and genital herpes (HSV-2) are much bigger particles with an icosahedral diameter of 200 nm, or one-fifth the size of a bacterium. HSV-1 and HSV-2 are dsDNA viruses whose genomes encode more than seventy genes. Seven different proteins are assembled to form the capsid and these are attached by other proteins to a surrounding lipid envelope. Following entry into the host cell, a shutoff protein encoded in the herpes genome stops synthesis of host proteins, destroys host mRNA, and regulates the expression of viral genes. HSV-1 and HSV-2 persist within cells of the nervous system as latent infections and become reactivated by a variety of environmental and physiological stimuli. This switching between latent and lytic (cell bursting) cycles accounts for the periodic outbreak of the symptoms of infection.

The application of the term bacteriophage to viruses that attack bacteria *and* archaea reflects a time when archaea were not recognized as a separate kind of prokaryote. A few of these viruses contain RNA and ssDNA, but the majority of bacteriophage genomes are formulated as dsDNA molecules. Bacteriophage morphology ranges from icosahedra, rods, and lemon-shaped particles to the elegant tailed viruses described earlier in this chapter (Figure 21). The dsDNA of these phages is packaged in an icosahedral head that connects to a tubular tail capped with a base plate surrounded by thin spider-leg fibres. The fibres flex when they make molecular contact with the surface of a bacterium, pulling the base plate onto the host cell wall. Binding of the plate to the wall causes contraction of the tail, which drives the viral DNA into the bacterium. The T4 phage utilizes this mechanism to infect *Escherichia coli* cells and served as one of the most important tools in early molecular biological research. T4 is a 200 nm-long virus and its genome encodes 289 proteins. Despite decades of work on this virus by teams of talented investigators, the function of many of its proteins is unknown.

In the last decade, microbiologists have discovered giant viruses whose genomes specify more proteins than some cells. Megaviruses, also known as large nucleocytoplasmic DNA viruses, replicate inside the cells of amoebae and other unicellular eukaryotes including planktonic algae. These extraordinary viruses match the dimensions and genetic endowment of many bacteria. These are large enough to be seen using a light microscope and investigators wondered, at first, if they were bacterial parasites residing in their hosts. Some of the

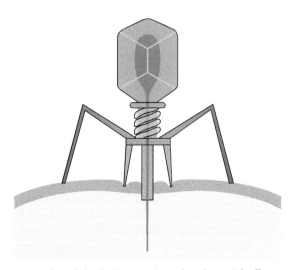

21. Bacteriophage injecting its DNA into a host bacterial cell

handful of sequenced genomes include genes that encode enzymes involved in DNA replication and gene expression, but all of the giant viruses remain dependent upon the metabolic activity of cells and their translational machinery. Pandoraviruses are the biggest of the viruses discovered so far. They occupy amoebae, forming ovoid particles surrounded by a lipid envelope, and lack the instructions for making capsid proteins. The origin of the megaviruses is an interesting topic with some investigators suggesting that they may have evolved from cellular ancestors through the loss of their physiological independence. Another suggestion is that the megaviruses represent one or more novel biological domains whose significance is comparable to the bacteria, archaea, and eukaryotes. The existence of megaviruses is an exciting and humbling discovery that shows how much we have to learn about biological diversity. Every prokaryote and eukaryote is colonized by multiple viruses and we have very little information on the way in which the majority of these infectious agents affect their biology.

Defective viruses have lost portions of their genome and must co-infect host cells with a helper virus to execute replication. They are capable of disabling the helper viruses by competing with their replication mechanisms, but instances of defective viruses increasing the transmission of the helper viruses have been found. P2 is a helper virus whose genome encodes the capsid protein for bacteriophage P4 (not T4 mentioned above), which attacks *Escherichia coli*. Successful infection of bacteria requires the transcription of proteins from P4 and P2, and the partnering viruses ensure mutual replication through this molecular cooperation. The Sputnik virophage is another kind of defective virus that co-infects amoebae with megaviruses. This virophage relies upon megavirus proteins to copy itself and causes the inhibition of megavirus replication. Sputnik can be regarded as a parasite of the giant virus.

Besides viruses, and the bacteria and fungi that cause disease (Chapter 5), a variety of simple molecular entities are responsible for highly destructive infections. Viroids are naked forms of infectious RNA that resemble viruses in their mechanism of replication. Lacking a protein capsid, these ssRNA molecules cause plant diseases by entering wounded tissue and migrating between cells to colonize the host. Viroids reduce the yield of potato, apples, avocado, aubergine (eggplant), and other agricultural crops. Replication of the viroid RNA is catalysed by RNA polymerase in the host plant. Interference with the expression of plant genes by the viroid seems to be the likeliest mechanism that results in damage to the plant tissues.

Prions are protein-based infectious agents of animals that do not contain any nucleic acid. Examples of prion illnesses include mad cow disease (bovine spongiform encephalopathy) and its lethal human manifestation, Creutzfeldt-Jakob disease (CJD). Prions cause disease by interacting with proteins encoded in the host genome, converting a harmless molecule into a damaging protein. The conversion process involves the misfolding of the host protein that causes cell death. In CJD the prion protein accumulates in brain tissue, killing nerve cells and perforating the brain with holes.

Chapter 5
Microbiology of human health and disease

Microbiological research has been dominated by studies on pathogenic organisms since the work of Louis Pasteur and other 19th century pioneers in medicine. The remarkable progress in understanding and treating infectious disease is the product of one of the most successful human endeavours in history. It is a triumph of the scientific culture of the industrialized Western world. An unintended consequence of these early investigations is that there is a common tendency to view *all* microorganisms as a pernicious component of life. This picture is beginning to change. Recent research suggests that populations of microbes that live in our digestive, reproductive, and respiratory tracts are as important to our wellbeing as the avoidance and treatment of infection. This molecular analysis of the human microbiome may stimulate a future revolution in medicine.

The average human comprises forty trillion eukaryotic cells and an accompanying microbiome of a hundred trillion bacteria, mostly in the gut, and one quadrillion viruses. We are, in raw cell numbers, more microbe than mammal. In addition to bacteria and viruses, the microbiome contains archaea, plus fungi, and other eukaryotic microorganisms. The majority of these microbes are beneficial; a minority have the potential to cause disease. Foetal exposure to microorganisms seems to be limited, although some studies indicate that a few bacteria may be ingested from the amniotic fluid before birth. During birth, the baby is coated with vaginal bacteria and this initiates our microbe-rich symbiosis. Babies born by Caesarian section tend to be coated with microbes from the skin of the mother and other adults that handle the newborn infant. These organisms seed the microbiome of the infant gut. Early colonizers include *Lactobacillus* and other firmicutes (Chapter 1), which are tolerant of acidic conditions, and these are replaced by species of bacteroidetes that decompose plant carbohydrates when a solid diet is introduced.

The prevalence of different species of bacteroidetes reflects the balance between plant carbohydrates that dominate diets in developing countries, and animal proteins and saturated fats that tend to be more common in Western diets. We rely upon microorganisms to digest the complex carbohydrates that form the cell walls of plant tissues because our genome encodes few of the necessary enzymes. Bacteria that process these polymers produce low molecular weight compounds including short chain fatty acids. Rupture of the bacteria, caused by natural cell ageing and viral infection, releases the fatty acids and other chemicals to the gut epithelium where they are metabolized. The importance of this symbiotic fuel supply is indicated in experiments with germ-free rats raised without a microbiome. These 'gnotobiotic' rats require a richer than normal diet to replace the calories delivered by the microbiome.

In addition to their nutritional value, metabolites produced by the microbiome confer other benefits including the regulation of water and salt absorption from the gut, the inhibition of pathogenic microbes, and maintenance of healthy levels of lipids and glucose in the bloodstream. Vitamin synthesis by gut bacteria is another feature of the microbiome that is critical for our survival.

Bacteria dominate the human gut and coexist with smaller populations of archaea and unicellular eukaryotes. Despite their relatively low numbers—one hundred million cells of archaea in a gram of faeces compared with forty billion bacteria—the minority residents are crucial players in digestive processes. Anaerobic archaea operate as methanogens, consuming hydrogen gas and producing methane. These microorganisms seem to grow close to the wall of the colon and cooperate with bacteria in the efficient fermentation of sugars. Some studies suggest that large populations of methanogens are associated with obesity and also become abundant in anorexic patients. This seeming contradiction may be explained by the common link between the digestive efficiency in obesity, resulting in weight gain, and in anorexia, when the microbiome is working hard to deliver enough calories under conditions of starvation. Other studies are equivocal about the link between methanogens and weight gain, but they all highlight the complexity of the symbiosis. Eukaryotes in the healthy gut microbiome include fungi and stramenopiles, whose role in human health is unclear, and known pathogens that are not causing active infections. Animal experiments show that fungi, particularly yeasts, multiply in response to gut inflammation and that this change in ecology can be linked to colitis.

The use of antibiotics to treat bacterial infections has significant and long-lasting effects upon the gut microbiome. A single course of antibiotics reduces the diversity of the gut bacteria and genes that confer resistance to the antibiotic spread rapidly among the survivors. Experiments show that these effects can persist for at least two years. This is an alarming finding given the growing body of research linking the composition of the gut microbiome to many illnesses. Inflammatory responses to gut microbes have been associated with ulcerative colitis, rheumatoid arthritis, multiple sclerosis, diabetes, dermatitis, and asthma. Evidence for a causal mechanism bridging an alteration in the microbiome to the development of these conditions is limited, but this field of clinical research is developing very swiftly. Widespread interest in improving health by altering the gut microbiome has created a huge market for foods containing live cultures of microorganisms. Evidence for the effectiveness of these probiotics is limited. The use of faecal transplants to reboot the gut microbiome of patients suffering from a variety of conditions is another therapeutic approach driven by interest in the relationship between gut microorganisms and our health.

Disturbance of the human microbiome in other locations can be harmful too. Mucous in the mouth is colonized by thousands of different bacterial species, plus archaea and yeasts. Dental plaque is a biofilm containing hundreds of species of microbes embedded in a polysaccharide matrix. The structural complexity and diversity of organisms in plaque biofilms is astonishing. A layer of rod-shaped bacteria attached to the tooth enamel is covered with spindle-shaped cells and the biofilm is capped with a mixture of filamentous, rod-shaped, and coccoid cells. A variety of bacteria, including spirochetes, and yeasts, forms a loose aggregate on top of the tight groupings of cells in the biofilm. Our nasal passages and lungs support other distinctive mixtures of bacteria and the skin is rich in yeasts.

Only a few of the rich assortment of microorganisms that populate the human microbiome are regarded as disease-causing agents. Microbiologists refer to the ability of a microbe to produce disease as its pathogenicity. The term virulence is often used interchangeably with pathogenicity, but sometimes specifies the degree of pathogenicity. Ebolaviruses, for example, are highly virulent, with fatality rates reaching 90 per cent in some outbreaks of haemorrhagic fever. Some microbes are obligate pathogens whose development is always associated with disease. Opportunistic pathogens can be constituents of the healthy microbiome and cause disease only under conditions when the host defences are weakened in some fashion. Many fungi are opportunistic pathogens that cause life-threatening infections in patients with impaired immune systems.

Infections are initiated by microorganisms that attach to our mucous membranes. Mucous membranes are the layers of epithelial cells that line internal organs and form a barrier against the external environment. These membranes secrete mucous in some locations. Microorganisms that become trapped in this sticky fluid are shed from the epithelial surface. Pathogens avoid removal by recognizing receptor proteins on the epithelial cells and adhering to the host cell surface. Adhesion provides microbes with an opportunity to invade the underlying tissues. Pathogenic microbes are introduced to the body via inhalation, physical contact with an infected host, bites from arthropods and other animal vectors, consumption of contaminated food and drinking water, and exposure to infectious microorganisms in soil.

Sneezing and coughing propel droplets of mucous loaded with bacteria and viruses into the air. Air expulsion during a sneeze is a particularly effective mechanism for microbial transmission, discharging tens of thousands of tiny droplets at speeds of up to 100 metres per second and misting the environment with microorganisms. Even the act of talking releases material from the onboard microbiome. Transport of microbes within mucous is important, because air-drying destroys most bacteria and viruses. Exceptions to this rule include bacteria that produce cell coatings that protect against drying. In general, Gram-positive bacteria with thick cell walls are more resistant to desiccation than Gram-negative species. The most lethal form of anthrax results from the inhalation of thick-walled endospores of *Bacillus anthracis* that survive heating and drying.

Familiar diseases transmitted by airborne bacteria include a variety of infections caused by *Streptococcus* species, diphtheria and pertussis (whooping cough), tuberculosis, and meningococcal meningitis. 'Strep throat', or streptococcal pharyngitis, is caused by *Streptococcus pyogenes* that lives as a benign member of the microbiome in the upper part of the respiratory tract. It operates as an opportunistic pathogen when host immune defences are weakened. In addition to sore throat and associated symptoms of this common infection, this bacterium causes middle ear infections, inflammation of breast tissue (mastitis), and skin disease. Necrotizing fasciitis is a severe condition caused by *Streptococcus pyogenes* in which the bacterium forms a deep-seated infection beneath a wound and causes widespread tissue damage that can result in death. This rare illness is referred to as the 'flesh-eating bacterium' in news reports.

Scarlet fever is a common streptococcal infection in children. The bright red skin rash is induced by a toxin that stimulates the accumulation of T cells of the immune system in the infection sites. Interestingly, the toxin is encoded by a gene (*speA*) carried by a bacteriophage (bacteriophage T12) and incorporated into the *Streptococcus* genome as a lysogenic infection. Infection of the bacterium with this virus results in the conversion of *Streptococcus pyogenes* from a non-virulent to a virulent strain. Scarlet fever usually fades after five or six days. A potentially fatal illness, called streptococcal toxic shock syndrome, is caused when the speA toxin activates cytokine secretion by a huge population of T cells. The resulting 'cytokine storm' causes massive inflammation and tissue damage. Infections by *Streptococcus pyogenes* are responsible for many other diseases, including rheumatic fever that can cause permanent damage to the heart, kidneys, and joints. A related species, *Streptococcus pneumoniae*, is responsible for half of all cases of pneumonia; many kinds of bacteria can cause this inflammatory lung disease and other microbial causes of pneumonia include fungi, unicellular eukaryotes, and viruses.

Tuberculosis (TB), caused by *Mycobacterium tuberculosis*, is a highly contagious illness spread from person to person by air. The majority of people exposed to the bacterium do not develop disease symptoms, but 10 per cent of these asymptomatic infections progress into an active disease. Most cases of TB occur in developing countries and more than one million people die from the illness every year. Among infectious agents, only HIV kills more people. In most active infections the bacterium damages lung tissue, but it spreads to other organs in a small proportion of cases. Damage to the immune system caused by HIV is a major risk factor for the development of TB. Another mycobacterium, *Mycobacterium leprae*, causes Hansen's disease or leprosy.

Influenza is an airborne infection caused by RNA viruses. Seasonal influenza occurs during cooler months every year, with the number of cases peaking in February in the United States and in August in Australia. The reason for this seasonality is unknown, but there is some evidence that transmission of the virus in mucous droplets is favoured in cold and dry air. Wild birds are a reservoir for influenza A that causes disease outbreaks in domesticated chickens and produces influenza pandemics in humans. Two types of glycoprotein cover the surface of this virus: hemagglutinin (H) binds to the surface of epithelial cells in the throat and lungs, and a second glycoprotein, neuraminidase (N), functions in the release of the replicated virus from infected cells. Different subtypes of influenza A are identified by their glycoproteins: e.g. H1N1, which caused the Spanish Flu pandemic in 1918–20 and Swine Flu in 2009, and H3N2, which caused Hong Kong Flu in the 1960s. The reason that the H1N1 virus caused 50 million or more deaths in the Spanish Flu pandemic is unclear, but it seems to have induced an inflammatory cytokine storm similar to a bacterial toxic shock syndrome. Other viruses transmitted by air cause measles, mumps, rubella, chickenpox and shingles, and the common cold.

Spores of pathogenic fungi are also spread by air. Aspergillosis is a lung infection caused by species of the ascomycete fungus *Aspergillus*. All of us inhale the spores of *Aspergillus*, but the competent immune system is very effective at preventing growth of these microorganisms. Suppression of the immune system by HIV infection or as a side effect of cancer therapy, as well as lung damage caused by tuberculosis, provides *Aspergillus* with an opportunity to develop further. The active fungus grows in lung tissues and can form an obstruction called an aspergilloma, or fungus ball, of compacted hyphae and dead lung tissue. In rare cases, *Aspergillus* hyphae can penetrate tissues beyond the lung, causing a fatal invasive form of aspergillosis. Because aspergillosis develops in response to a compromised immune system or tissue damage it is categorized as an opportunistic infection.

Opportunistic infection is characteristic of most fungal diseases. *Histoplasma capsulatum* and *Blastomyces dermatitidis* cause pneumonia symptoms and a wide range of illnesses described by the generic terms histoplasmosis and blastomycosis. *Histoplasma* grows in bird and bat guano and its spores are inhaled deep into the lung and reach the alveoli. The spores survive ingestion by the macrophages of the immune system and can spread to other parts of the body through the lymphatic system. Histoplasmosis, known as Ohio Valley disease, is endemic to the Ohio River Valley and Lower Mississippi River. Cryptococcosis, caused by *Cryptococcus neoformans* and *Cryptococcus gattii*, originates as a lung infection and progresses as a disease of the central nervous system and causes meningitis. It is most common as an opportunistic infection in AIDS patients. Coccidioidomycosis or valley fever, caused by *Coccidioides immitis*, is a chronic pneumonia that can become a lethal infection. The fungus lives in desert soils in Arizona, California, northern Mexico, and parts of Central and South America. *Coccidioides* spores become airborne when dry soil is disturbed by rainstorms, earthquakes, and construction projects. Coccidioidomycosis is unusually virulent for a fungus. It infects patients with healthy immune systems, but most cases resolve without treatment. Zygomycetes that flourish on food scraps cause rare infections of the nasal sinuses called mucormycosis that can spread to the brain. Even mushroom-forming fungi have been reported as disease agents in a few clinical cases.

Unlike so many infections caused by bacteria and viruses, none of the opportunistic infections caused by fungi are contagious. Dermatophytes are pathogenic fungi that grow on the skin, hair, and nails, causing ringworm and athlete's foot. The fungi are not considered to be opportunists and they can be contagious, spreading through physical contact and shared use of towels and clothing contaminated with spores. Dermatophytes are filamentous ascomycetes, including species of *Trichosporon* and *Microsporon*, which feed on keratin. In most cases, only limited areas of tissue are colonized by these fungi, but, in some instances, dermatophytes can spread and cause more serious illnesses. *Malassezia* species are part of the normal microbiome of the scalp. These yeasts feed on lipids in the skin secretion called sebum. *Malassezia globosa* causes dandruff and can spread and cause an inflammatory infection called seborrhoeic dermatitis.

Transmission of many bacterial and viral pathogens requires direct physical contact with an infected individual or contaminated blood. *Staphylococcus aureus* is a component of the normal microbiome and virulent strains of the bacterium cause skin diseases, pneumonia, meningitis, and other illnesses. Healthy people can be carriers of the bacterium and there is a lot of concern about the spread in hospitals of methicillin-resistant strains (MRSA) that are impervious to conventional antibiotic treatment. Hepatitis A is transmitted in water and food contaminated by faeces, and hepatitis B spreads via contact with blood containing the active virus. Gonorrhoea, syphilis, and chlamydia are sexually transmitted diseases (STDs) caused by bacteria; AIDS, herpes, and human papillomavirus infections are STDs resulting from viral infection.

The symptoms of infectious diseases are caused by tissue damage, toxin production by some bacteria, and the inflammatory response of the immune system. HIV is an example of a more complex infection whose debilitating effects are due to opportunists whose growth is fostered by the viral destruction of macrophages and T cells in the immune system. Common secondary infections include fungal pneumonia caused by *Pneumocystis jiroveci*, brain lesions caused by *Cryptococcus neoformans*, and disseminated candidiasis caused by *Candida albicans*. The immune impairment resulting from replication of HIV also promotes illnesses caused by parasitic protists including cryptosporidiosis (severe diarrhoea) and toxoplasmosis (brain inflammation). Both of these parasites—*Cryptosporidium* and *Toxoplasma*—are alveolates (Chapter 1). Another AIDS-defining illness is Kaposi's sarcoma, a tumour caused by a herpesvirus, HHV-8.

Pathogens spread by animals include rabies virus from animal bites, and a range of haemorrhagic fevers including hantavirus infections from exposure to buildings infested with rodents, and Ebola that may be acquired from bushmeat (as well as infected human tissue). Arthropod vectors spread plague (flea/bacterium *Yersinia pestis*), West Nile fever (mosquito/West Nile virus), Lyme disease (tick/bacterium *Borrelia burgdorferi*), malaria (mosquito/ alveolate protist *Plasmodium*), and typhus (louse/bacterium *Rickettsia*). The scale of suffering caused by these diseases is staggering. Plague, as the presumed cause of the Black Death, killed 75 to 200 million people in the 14th century; typhus decimated populations of prisoners in concentration camps and refugees during the Second World War, and malaria continues to massacre half a million people every year.

Water and food contaminated with faecal microorganisms are another source of microbial illnesses. Cholera (bacterium *Vibrio cholerae*), typhoid (bacterium *Salmonella enterica*), and viruses, and amoebiasis (amoebozoan *Entamoeba histolytica*) are major waterborne illnesses; bacterial toxins produced by *Staphylococcus aureus*, *Escherichia coli*, and *Salmonella* (salmonellosis) cause food poisoning. Botulinum toxin is a lethal neurotoxin produced in processed foods when endospores of *Clostridium botulinum* germinate and grow under anaerobic conditions. Foodborne botulism is quite rare today. The related tetanus bacterium, *Clostridium tetani*, proliferates in wounds that present an anaerobic environment. It produces a toxin called tetanospasmin that interferes with nerve impulses, eliciting muscle spasms that can lead to respiratory failure.

Contrary to the impression that may be left by this lengthy catalogue of infectious agents, relatively few organisms within the great sweep of microbial diversity cause disease in humans. Bacteria and viruses are the most important disease agents, but most of them are harmless toward humans. Opportunistic fungal infections can be lethal, but only a handful of the tens of thousands of described species of fungi cause most of the problems. A similar minority of single-celled protists, mostly within the alveolate supergrouping, are pathogenic. And there are no examples of diseases caused by archaea alone. Some studies indicate that archaea may contribute to tissue damage caused by communities of prokaryotes. Tooth decay is an example of this kind of polymicrobial infection.

Many of the microorganisms that cause opportunistic infections seem poorly adapted for growth in animal hosts: they can grow at the elevated temperatures presented by the body and meet their energetic needs by digesting tissues, but they are at the mercy of immune defences that normally exclude them from the animal. Hosts with impaired defences certainly offer a food source for opportunists, but the low levels of iron and other micronutrients in body fluids are unwavering obstacles to microbial growth. The suggestion that the animal host is a 'dead end' for the pathogen is not correct, however, because opportunists in nature are returned to the environment when the body decays and can then locate the food sources that support them outside the host.

Defences against pathogens are categorized as innate mechanisms, which are 'built in' and do not require prior exposure to the microorganism, and adaptive mechanisms that operate when the body reacts to the recognition of foreign proteins. White blood cells called macrophages and neutrophils are responsible for innate immunity. They are programmed to detect molecules described as pathogen-associated molecular patterns (PAMPs) that are characteristic of groups of microorganisms rather than individual species. PAMPs include large molecules called lipopolysaccharides that are found in the outer membranes of all Gram-negative bacteria, the protein flagellin that forms the filament of the bacterial flagellum, and common components of cell walls. PAMPs are detected by receptors on the surface of macrophages that destroy the foreign cells by phagocytosis. In adaptive immunity, cells of the immune system destroy pathogens and present antigens from these cells or viral particles—usually proteins—to T lymphocytes. Antigen presentation stimulates T cell differentiation into cytotoxic T cells (killer T cells) that destroy host cells presenting the antigens, or into helper T cells (T_H1) that release cytokines and cause an inflammatory response that limits spread of the infection. A second type of helper T cell, T_H2, induces B cell differentiation that results in antibody production. Antibodies or immunoglobulins are glycoproteins that bind to single antigens on the surface of pathogens, neutralizing the invading cells or tagging them for destruction by other immune cells.

When our immune defences are unable to combat bacterial infections they can be treated with antibiotics. These compounds work by attacking cell wall synthesis, or they target other cellular mechanisms including transcription of DNA into messenger RNA and translation of mRNA into protein. Some antibiotics are effective against Gram-negative and Gram-positive bacteria, others have a narrower spectrum of activity. Penicillins, secreted by species of the ascomycete fungus *Penicillium*, are powerful antibacterial agents that target Gram-positive bacteria. They are members of a class of antibiotics that share a core structure called the β-lactam ring. Synthesis of penicillin G (Gold Standard penicillin) by *Penicillium chrysogenum* was discovered by

Alexander Fleming in 1928 and developed for clinical use by a research team of Oxford University scientists led by Howard Florey. Research to increase production of the antibiotic began in 1938 and the drug was tested on the first patient in 1941. A different strain of *Penicillium chrysogenum*, isolated from a rotting cantaloupe, gave much higher yields of the antibiotic and this fungus became the mainstay of industrial production. Penicillins and other β-lactam antibiotics destroy bacterial cells by interfering with the formation of cross-links within their peptidoglycan cell wall. Inhibition of these enzymes weakens the cell wall and causes the bacteria to burst.

Bacterial resistance to penicillin derives from the evolution of strains that produce enzymes called β-lactamases that inactivate the antibiotic by opening its β-lactam ring. Bacterial resistance became a problem within the first decade of penicillin's use. An important advance was made with the discovery of the core structure of penicillin, a molecule called 6-aminopenicillanic acid, produced by *Pencillium chrysogenum*. This molecule lacks antibiotic activity, but can be used as a starting material to produce semi-synthetic antibiotics by adding a variety of side chains. Methicillin (meticillin) and ampicillin are examples of these compounds. Unlike penicillins, ampicillin is a broad-spectrum antibiotic, active against Gram-negative as well as Gram-positive bacteria.

Cephalosporins are β-lactam antibiotics produced by the fungus *Acremonium chrysogenum*. The first generation of cephalosporins were active against Gram-positive bacteria. Later versions of these antibiotics, produced by cleaving the parent molecules and adding various side chains, have shown greater efficacy against Gram-negative bacteria, sometimes at the expense of reduced potency against Gram-positive bacteria. The latest fifth-generation cephalosporins are exceedingly valuable antibiotics representing the last line of defence against infections caused by methicillin-resistant *Staphylococcus aureus* (MRSA).

Streptomycin, erythromycin, and tetracyclines are powerful antibiotics produced by the filamentous bacterium *Streptomyces*. Erythromycin is used in patients that are allergic to penicillin. *Streptomyces* is also a source of polyenes that are used to treat fungal infections. Polyenes bind to ergosterol, which is a component of the plasma membrane of fungi, making the cells leaky. Ergosterol is absent from animal cell membranes, making it a good target for antifungal drug therapy. Polyene antifungals, including amphotericin B, are not perfect, however, because they also attack animal membranes causing kidney damage and other side effects. Host toxicity is also characteristic of most antiviral agents because these compounds work by disrupting the machinery for viral replication within the infected cells. Nucleoside analogues inhibit polymerases, reverse transcriptases, and other enzymes involved in the replication of the viral genome. Toxicity of these compounds toward the host occurs because normal replication of the host DNA is affected. Protease inhibitors have proven very effective treatments for HIV. These bind to protease enzymes that function in capsid assembly. None of the antibacterial antibiotics has any activity against viruses.

Respiratory allergy is a distinctive type of immune response triggered by antigens carried on the surface of airborne particles, including fungal spores, pollen, and pet dander. It is significant from a microbiological perspective because millions of tons of fungal spores are released into the air every year and are a major contributor to asthma and allergic rhinitis. Antigens on the surface of these spores are processed by dendritic cells in the lung and presented to helper T cells (T_H2 rather than T_H1 cells) that stimulate antibody production by B cells. Immunoglobulin E (IgE) is the class of antibody produced in response to this hypersensitivity reaction and these glycoproteins bind to receptors on the surface of mast cells. Mast cells bearing the IgE molecules are sensitized to the allergen and when they encounter the same antigen again they release histamine and other inflammatory compounds. This degranulation process increases blood flow and blood vessel permeability, producing the symptoms of asthma and rhinitis. Allergies are a serious global health problem, with an estimated 300 million people suffering from asthma.

There has been a lot of concern, particularly in the United States, about the purported toxicity of certain fungi that grow in flooded homes. The spores of some of these indoor fungi, including a black-pigmented ascomycete called *Stachybotrys chartarum*, carry toxins that can cause a range of illnesses if they are absorbed in high concentrations. Most of the available evidence suggests that people that inhale spores of this fungus in water-damaged buildings are not exposed to levels of these mycotoxins that can cause illness. Nevertheless, the inhalation of large quantities of allergenic spores in these circumstances remains a serious public health concern.

Chapter 6
Microbial ecology and evolution

All biology was microbiology for 2.6 billion years before the evolution of multicellular organisms. Many ecosystems remain wholly microbial and microorganisms continue to control environments that appear to be governed by plants and animals. Earth is an exceedingly microbiological planet. And, if life evolves elsewhere in the universe, microbes rather than larger organisms are certain to be the most numerous residents. Reason for confidence in this assertion comes from considering the flow of energy in our biosphere.

The activities of microorganisms provide the biochemical foundation for plant and animal life. This may not be obvious, immediately, because we know that solar energy supports most life through photosynthesis and plants are the major primary producers on land. Plants dominate productive land areas, outweighing animals by a factor of 1,000, but their total mass is probably matched by prokaryotes. Furthermore, prokaryotes account for a far greater proportion of the living biomass because most of the carbon in plants is sequestered in cell wall polymers that constitute wood.

Despite their autotrophic lifestyle, plants are not self-supporting organisms. They depend upon the complex redox reactions of microbes that fertilize the soil by fixing nitrogen, converting nitrites to nitrates, enhancing the availability of phosphorus and trace elements, and recycling organic matter (Chapter 2). Eukaryotic microorganisms are also plentiful and essential for the sustenance of plants and animals. Mycorrhizal fungi supply plants with nutrients and water and improve soil structure. Saprotrophic fungi, amoebozoans, and ciliated protists compete for resources and drive carbon and nitrogen cycling in association with the more numerous prokaryotes. Plants and the animals that eat them rely on microorganisms in the soil.

The significance of microorganisms in energy flow is more apparent in the ocean where plants are absent. Cyanobacteria and single-celled eukaryotic algae are the most active photosynthetic organisms in seawater. Photosynthetic and non-photosynthetic microbes are the masters of the marine environment, harnessing the energy that supports complex ecological interactions between aquatic animals. Bacteria and archaea form 90 per cent of the ocean biomass and surface waters are filled with eukaryotic algae. Marine cyanobacteria absorb as much carbon dioxide as all of the tropical rainforests, and planktonic diatoms capture an equivalent amount of carbon. Together, the photosynthetic microorganisms of the seas match the photosynthetic activity of plants on land. Viruses, particularly bacteriophages, are the most abundant genetic entities in the ocean. Rough estimates suggest that the lysis of cells by marine viruses removes 20 to 40 per cent of all of the bacteria every day. This astonishing viral turnover of planktonic microbes fertilizes the water below the sunlit surface of the sea.

Chemosynthetic microorganisms support flourishing ecosystems around hydrothermal vents and cold seeps on the seafloor and ancient populations of prokaryotes and fungi grow in abyssal sediments to depths of hundreds of metres. Bacteria may colonize the Earth's crust to depths of a few kilometres before they are extinguished by heat from the mantle. The biomass of this hidden microbiome, called the Deep Biosphere, is unknown but could, conceivably, outweigh all of the organisms above. This controversial idea highlights how much we have to learn about microbial ecology. It also expands the range of planetary conditions that may be capable of supporting life and makes it more probable that microorganisms have evolved elsewhere in our solar system and beyond.

Soil is a more complex habitat for microorganisms than aquatic eco-systems because it has a three-dimensional structure and chemistry that varies over distances ranging from microscopic to geographic. The physical structure of soil is defined by the proportions of sand grains, tiny fragments of silt, and clay particles that are even smaller than bacteria. Sandy soils tend to be well aerated but less fertile than those with a greater percentage of silt and clay. Clay has a crystalline structure that attracts ions from percolating water and can exchange them with plant roots. Because clay particles are so small, they furnish an enormous surface area for these chemical reactions: a cubic metre of clay has a surface area of six million square metres. Soils are enriched by the fragmentation and decomposition of plant and animal tissues by invertebrates and microorganisms. Some of the organic matter is highly insoluble and forms humus that resists further breakdown for hundreds or thousands of years. This humus affects soil structure, moisture content, and nutrient retention.

Another level of ecological complexity is introduced when we consider that each soil bacterium has a unique life experience shaped by fluctuations in the availability of dissolved ions, changes in temperature, contact with other bacteria, and attack by viruses. Bacteria have some control over their fate. Gene expression inside the tiny cells is adjusted continuously to maintain energy production and maximize the prospects for cell division. Motile cells with spinning flagella navigate through films of water surrounding soil particles, responding to gradients of dissolved oxygen and organic nutrients as well as local clouds of metabolites secreted by other organisms. Non-motile bacteria are stuck in the sites colonized by their 'parental' clones unless they are displaced by the percolation of water or hitch a ride on a nematode worm.

Every pinch of soil is a mosaic of microbes. Sweeping descriptions of the effects of microorganisms on soil chemistry do not begin to capture the ecological complexity of the habitat. The same gap between the overall effects of microbial communities on their surroundings and the details of chemistry on the microscopic scale applies to the oceans and every other habitat. We are unaccustomed to thinking about the spatial scale of the environment that matters to single cells, but this is an important consideration in microbial ecology.

One gram of rich forest soil contains an estimated one hundred million prokaryotes and metagenomic studies reveal thousands of different kinds of bacteria in samples from single locations. It is tempting to call these genetically distinctive organisms species, but we know so little about these bacteria that we use the terms 'phylotype' and 'operational taxonomic unit'. These refer to genetic sequences that show sufficient differences that they appear to come from populations of microorganisms that may represent different species. Part of the uncertainty comes from difficulties in defining species of microorganisms. Biologists consider reproductive incompatibility for separating animal species, but the 'rules' for zoology and botany are ineffective for prokaryotes that do not reproduce sexually and for most eukaryotic microbes too. The nature of microbial species is complicated further by the fact that the majority of soil microbes identified using molecular techniques cannot be cultured. This means that almost nothing is known about their biology. Nevertheless, the diversity of soil microbes is remarkable and shows huge variations between soils in different locations and with different properties.

The abundance of soil viruses eclipses the number of cellular microorganisms and bacteriophages show seemingly limitless genetic diversity. It is estimated that each species of bacterium—in the soil and in other habitats—serves as prey for ten or more kinds of bacteriophage. If this is correct, the estimated ten million species of bacterium may support one hundred million kinds of phage. The application of the term 'species' is even more problematic for viruses than bacteria and we rely upon arbitrary measures of genetic novelty when we refer to different kinds of bacteriophage. Sequencing studies show tremendous genetic variation among viruses that infect bacteria and the ease with which 'new' viral genes are sequenced suggest that billions more await discovery. Viruses that infect archaea and eukaryotes also represent immense repositories of genetic information.

Proteobacteria are the commonest soil bacteria, followed by acidobacteria and bacteroidetes. Soil proteobacteria include photosynthetic purple bacteria, pseudomonads, and myxobacteria. *Rhizobium* species are proteobacteria that form root nodules with legumes and produce ammonia from atmospheric nitrogen, and nitrifying proteobacteria convert ammonia to nitrite and nitrite to nitrate. *Rhizobium* provides its host with ammonia and amino acids and is fed with sugars from the plant. The nitrifying bacteria are 'free-living' soil microbes that meet their energy needs by oxidizing ammonia and nitrite and use the electrons from these reactions to form sugars by reducing carbon dioxide. Nitrifying bacteria are chemolithotrophs. Nitrifying archaea that oxidize ammonia are more abundant in soils than bacteria that perform this reaction. The genetic signature for this process is the gene (*amoA*) that encodes the enzyme ammonia monoxygenase. The return of nitrogen to the atmosphere is accomplished by *Pseudomonas* and other bacteria that convert nitrate to N_2. This denitrification mechanism occurs under anaerobic conditions in deeper soil where the bacteria use nitrate ions, rather than oxygen, as the terminal electron acceptor for their respiratory reactions.

Soil fungi show none of the metabolic variety of prokaryotes: all of them are chemoorganotrophs that consume organic materials produced by other organisms. The sources of these foods are very diverse, however, with fungi preying on bacteria, nematode worms, plants, and other soil inhabitants, decomposing herbaceous and woody plant tissues, and obtaining nutritional support from plants with which they form mycorrhizal symbioses. Structural diversity among soil fungi ranges from simple, single-celled organisms whose cells swim in moist soils, filamentous fungi that dissolve scraps of organic matter, and mushrooms that form monstrous colonies spanning thousands of hectares of forested land.

Fungi form different types of mycorrhizal relationship with plant root systems (Figure 22). Ectomycorrhizal partnerships involve mushroom-forming fungi and the roots of trees and shrubs. These symbioses support the dominant tree species in temperate and boreal forests. Fungal hyphae surround the affected roots producing a dense mantle and penetrate the walls between adjacent root cells, elaborating an interface for the exchange of water and dissolved nutrients. Colonies of mycorrhizal fungi can connect with multiple plants of the same species and also establish webs between different species. These 'common mycorrhizal networks' shape the succession of plants in forest ecosystems and affect the productivity of these complex communities.

A second kind of interface is produced by arbuscular mycorrhizal fungi that penetrate the roots of 90 per cent of land plants. Arbuscules are finely branched cells that extend into the cortical cells of roots, indenting the plant membrane, but not piercing it, establishing a live connection between the two partners. By ramifying through the surrounding soil, the filamentous hyphae of mycorrhizal fungi produce an accessory absorptive network that explores a much greater soil volume than the roots themselves and greatly increases the surface area for mineral uptake. The hyphae soak up phosphorus and other nutrients from the soil, supporting plant productivity in return for sugars that the host produces by photosynthesis. Other types of mycorrhizal symbiosis are formed between fungi and particular groups of flowering plants including

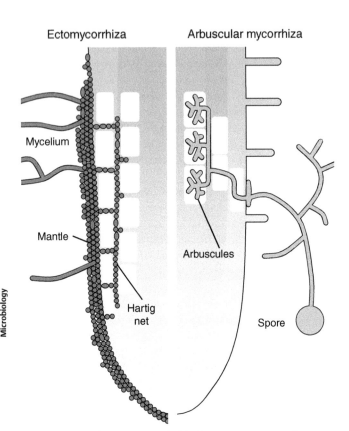

22. **Mycorrhizal fungi associated with plant root systems. Left, ectomycorrhizal fungi form a mantle on the outside of the root and grow between the root cells to form a structure called the Hartig net. Right, arbuscular mycorrhizal fungi form branched extensions called arbuscules inside root cells**

the orchids. Endophytic fungi and bacteria that live inside plant tissues form another kind of mutualism that is a critical player in terrestrial ecology. Endophytes secrete growth promoters and shield plants against pathogens.

Every animal participates in symbioses with microorganisms. Insects accommodate bacteria and eukaryotic microorganisms in their gut, support harmless microbes on the surface of their exoskeleton, and are prey to fungal, bacterial, and viral pathogens. All of these interactions are referred to as symbioses, using the broadest sense of the term. In colloquial usage, however, symbiosis is reserved for mutually supportive relationships or mutualisms.

Mutualistic symbioses include the extraordinary gut microbiome of termites and the partnership between leaf-cutter ants and their cultivated mushrooms. Termites are reliant upon *Trichonympha* and related anaerobic protists in the excavate supergrouping of eukaryotes for digestion of cellulose that they consume in the form of wood or grasses. Cellulose breakdown by *Trichonympha* is accomplished by thousands of endosymbiotic bacteria concentrated inside the bulbous end of each pear-shaped cell. The anterior of the *Trichonympha* cell is swathed in a swirling wig of 'ectosymbiotic' spirochete bacteria that operate as surrogate cilia. These bacteria, along with unattached spirochetes that populate the gut, participate in cellulose breakdown to form acetate. Acetate is absorbed by the insect gut and is the principal energy source for the termites. The gut microbiome of termites contains many other kinds of microorganisms including nitrogen-fixing bacteria, lactic acid bacteria, and archaea and bacteria that produce methane.

A different nutritional strategy has evolved among a group of mushroom-growing termites in Africa and elsewhere in the Old World that use fungi rather than gut protists for cellulose digestion. Termites grow *Termitomyces* mushrooms on their cellulose-rich faeces in towering mud mounds and eat the resulting compost. From South America to parts of the southern United States, leaf-cutter ants cultivate different kinds of mushrooms in their underground nests. The ants eat special buds that form as swellings on the fungal colony. They support another mutualism with actinobacteria that grow on the surface of their cuticle and secrete antibiotics that inhibit parasitic fungi that would otherwise destroy the fungus gardens.

Termite and ant symbioses with microorganisms are fascinating examples of mutualisms, but similarly complex relationships are crucial for all insects. Like termites, wood-eating cockroaches harbour anaerobic protists containing bacteria that cooperate in the process of cellulose digestion. A group of fungi called trichomycetes grow in the guts of insects, crustaceans, and myriapods (millipedes and centipedes). Little is known about the nature of these symbioses, but the prevalence of these fungi suggests that they are important members of the gut microbiome of their hosts. By highlighting a few classic cases of mutualism, it is easy to miss the fact that all multicellular organisms engage in symbioses with microorganisms. Nothing subsists as an individual.

Rivers, ponds, and other freshwater habitats support much lower numbers of microorganisms than soils: one million cells per millilitre is an ordinary tally for river water, compared with a hundred million cells in the same volume of soil. Primary productivity in these ecosystems is driven by plants, photosynthetic cyanobacteria, and eukaryotic algae. The phytoplankton includes cryptomonads, diatoms, dinoflagellates, euglenoids, and thousands of species of green algae that range from unicellular flagellates, to beautiful colonial algae like *Volvox* and elegant desmids shaped like needles, stars, and snowflakes. Carbon fixation by these organisms supports the heterotrophic microorganisms, invertebrates, and vertebrates in freshwater habitats through complicated food webs.

The mixture of planktonic organisms and their productivity are controlled by oxygen availability and temperature. As a lake warms in temperate regions, the density of the water falls and the upper part of the water column becomes separated from the colder and denser water below by a boundary called a thermocline. If the water is undisturbed, the oxygen levels underneath the thermocline plummet, supporting the multiplication of anaerobic bacteria and archaea. Later in the year, the different layers mix as the surface waters cool, the anoxic zone is suffused with oxygen, and the layering of aerobes over anaerobes is disrupted for a few months. The cycling of nitrogen in rivers and lakes involves many of the microbial processes described for soils including nitrogen fixation by cyanobacteria and the sequential formation of nitrite and nitrate by nitrifying bacteria. Denitrifying bacteria engage in the opposite reactions, reducing dissolved nitrate to N_2. In addition to this conventional form of denitrification, nitrogen gas is released by anaerobic bacteria that oxidize ammonia. These 'annamox' bacteria are active in soils and freshwater environments as well as the sea.

The biological significance of nitrogen is evident from its contribution to the structure of proteins and nucleic acids. Many other elements are essential components of biomolecules and are absorbed, transformed, and released by microorganisms. The maintenance of the biosphere is contingent upon this biological cycling of sulfur, iron, calcium, phosphorus, and silica. Sulfur, which is a constituent of proteins, is concentrated in the form of minerals in rocks. Weathering of these materials on land leads to the flow of sulfate (SO_4^{2-}) into the oceans and volcanic activity and hot springs release additional sulfur to the biosphere in the form of hydrogen sulfide (H_2S) and sulfur dioxide (SO_2). Emissions of SO_2 from burning fossil fuels have eclipsed volcanic sulfur emissions. Sulfate-reducing bacteria form H_2S in terrestrial and marine environments. These anaerobic organisms oxidize organic compounds or H_2 and use sulfate rather than oxygen as their electron acceptor. Their activity is responsible for the smell of rotten eggs in marshes and mud flats. Other bacteria and archaea reduce elemental sulfur (S^0) rather than sulfate to form H_2S. Some of these microbes flourish in the extreme environments associated with hydrothermal vents. Driving the reactions in the opposite direction, chemolithotrophic bacteria and green and purple photosynthetic bacteria power carbon fixation by oxidizing S^0 and H_2S.

Iron is cycled through a similar loop of oxidation and reduction reactions catalysed by bacteria and archaea. Iron-reducing microbes use the ferric form of the element (Fe^{3+}) as an electron acceptor under anaerobic conditions. This reaction occurs in waterlogged soils and other low oxygen environments. In transition zones where water from these locations trickles into an oxygen-rich environment, the dissolved ferrous iron supplies iron-oxidizing bacteria with their energy source. Bacterial oxidation of ferrous iron is favoured in acidic habitats including the drainage from coal mines. The resulting ferric oxide is insoluble and forms an orange-brown slime.

Other elements, including silica and calcium, are cycled through microorganisms without changing their redox state. Silica forms the shells, or exoskeletons, of diatoms, radiolarians, and other planktonic microbes. The remains of these organisms fall through the water as a component of a 'marine snow', the shells dissolving or accumulating on the seafloor in the form of siliceous ooze. Geological deposits called diatomaceous earth are evidence of the abundance of diatoms in ancient lakes and marine ecosystems.

The assembly of calcareous scales on the cell surface of coccolithophorid algae (members of the hacrobian supergroup) is another example of a microbial transformation that crystallizes a dissolved element into a hard cell structure (Figure 23). The algae absorb calcium and bicarbonate ions from seawater and fashion scales of calcite ($CaCO_3$) that they shift to the cell surface. This has important consequences for atmospheric chemistry, because the bicarbonate used by algae comes from carbon dioxide that dissolves in seawater. The formation of coccolithophorid scales is a sink for carbon dioxide. When these algae die, they settle through the water column and accumulate on the seafloor. This helps to maintain the slightly alkaline pH of seawater and reduces the concentration of CO_2 in the atmosphere. The magnitude of this process over the course of millions of years is evident from the fact that Britain's White Cliffs of Dover are formed from the

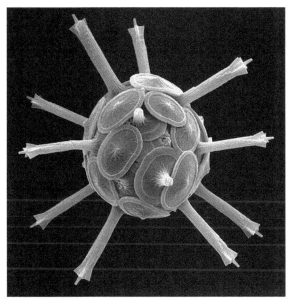

23. Electron micrograph of a marine coccolithophorid alga armoured with calcareous scales called coccoliths

scales of algae that bloomed and died in the Cretaceous. Coccolithophorids also affect atmospheric chemistry by releasing dimethyl sulfide (DMS) that acts as a seed for cloud formation.

Microorganisms have a huge influence on the chemistry of the atmosphere and have some ability to ameliorate the effects of atmospheric pollution by industrialized nations. Decomposition by bacteria, archaea, and fungi is the greatest source of CO_2 emissions, far exceeding the release of CO_2 from fossil fuel consumption. Natural CO_2 emissions are balanced by the absorption of CO_2 by photosynthetic and chemolithotrophic microorganisms as well as land plants. The concentration of CO_2 has been relatively stable for millions of years, but burning of coal, oil, and gas is upsetting this equilibrium, leading to a net increase in atmospheric CO_2 and consequent warming of the biosphere. Thirty to 40 per cent of the gas released by these human activities dissolves in oceans, lakes, and rivers, but there are limits to the operation of this natural sponge. Seawater is acidifying as it absorbs more and more CO_2 and this has profound consequences for all manner of marine life. Acidification and ocean warming affect scale formation by coccolithophorid algae and contribute to the expulsion of the dinoflagellates (zooxanthellae) whose photosynthetic activity supports corals. Research on microbial ecology is fundamental to our understanding of the causes and consequences climate change.

Microorganisms, particularly bacteria and archaea, thrive in environments that are intolerable for other organisms. These prokaryotes, along with a variety of yeasts and filamentous fungi, are called extremophiles. Physical conditions in the locations that prohibit other forms of life include high and low temperatures (thermophiles and psychrophiles), extremes in acidity and alkalinity (acidophiles and alkaliphiles), high salt (halophiles), low water availability (xerophiles), and lethal levels of ionizing radiation (radiotolerant microbes). A species of archaea, called *Pyrolobus fumarii*, grows at 113°C (235°F) in the chimneys of a hydrothermal vent along the Mid-Atlantic Ridge. It is 'chilled' at temperatures below 90° and stops cell division. Another unnamed archaea appears to grow at 121°C, which is the temperature of laboratory autoclaves used to sterilize instruments and glassware. Enzyme structure in these prokaryotes is modified to maintain catalytic function at higher temperatures and they have evolved mechanisms that facilitate quick refolding of damaged proteins.

At the other end of the temperature scale an archaea called *Methanogenium frigidum* lives at the bottom of an Antarctic lake where it can grow at 1–2°C in water saturated with methane. Protein function is conserved in this prokaryote through modifications that preserve flexibility of the molecular structure at low temperature. Cells of *Methanogenium* divide once per month, compared with a doubling time of twenty minutes for some of the bacteria in our gut. Cold marine environments teem with microorganisms inside the briny veins that permeate sea ice and in the cold sediments of the abyssal plain.

Natural pools of molten tar are among the stranger habitats that support bacteria and archaea. At Pitch Lake in Trinidad, oil from a deep deposit is forced to the surface forming a viscous tar that bubbles with methane. The density of prokaryotes in this liquid asphalt desert is comparable with forest soil, which is incongruous given that there is so little liquid water here. It seems likely that the bacteria and archaea inhabit brine-filled pockets in the tar, rather like the microbes that live in sea ice.

Archaea have been isolated from geothermal springs with acidity levels matching car batteries and from salt-saturated soda lakes with alkalinity equivalent to household ammonia. The acid tolerant cells have enhanced mechanisms for expelling protons from their cytoplasm. The membrane chemistry of alkaliphiles works in the opposite direction to import protons from a proton-deficient habitat. Salt lakes and bare rocks challenge microorganisms in a similar way by causing the dehydration of the cytoplasm. Adaptive mechanisms among halophiles and xerophiles include thickening of the cell wall to resist water loss, sugar and salt accumulation to subvert osmotic dehydration, and changes in cell shape to increase the surface area for ion exchange and water uptake. Radiotolerant bacteria withstand levels of gamma irradiation more than one thousand times greater than the lethal doses for humans. Their survival is attributed to the synthesis of multiple copies of the same chromosome in a single cell and use of uncorrupted sequences as templates for the repair of damaged genes.

The discovery of microorganisms in extreme environments has encouraged speculation about the evolution of life elsewhere in our solar system. There is no evidence for extraterrestrial life, but places where microorganisms similar to our extremophiles might exist include Saturn's moons: Titan, where the surface is showered with methane rain; and Enceladus, where cryovolcanoes erupt ice and water vapour. The freezing subsurface ocean on Jupiter's moon, Europa, is another conceivable habitat. The possibility that life has evolved beyond Earth—in our solar system and farther away on Goldilocks planets associated with other stars—is treated by some investigators as a starting point for investigations on extraterrestrial origins for life here. The discovery of extraterrestrial life would certainly advance the possibility of an alien seed for our existence but would do little to answer the greatest unsolved question in biology: How does life begin anywhere?

The most compelling ideas about the origin of cells come from the study of the chemistry and microbiology of hydrothermal vents. Heated water spews from seafloor vents where the Earth's crust is fractured by the separation of tectonic plates along the mid-ocean ridges of the Atlantic and Pacific. Chimneys form above some vents where minerals dissolved in the hot water precipitate as they are mixed with cold seawater. Vents called black smokers release dissolved sulfides and white smokers emit somewhat cooler water rich in calcium, barium, and silicon. In the absence of sunlight, vent chemistry supports chemolithotrophic prokaryotes that synthesize sugars by oxidizing hydrogen sulfide and other geofuels including hydrogen, methane, ammonia, ferrous iron, and manganese. (This is analogous to the photosynthetic capture of CO_2 powered by electrons obtained from water and energized by light absorption.) According to one model, the discharge of alkaline water through the porous chimneys of white smokers would have created a natural pH gradient in ancient oceans that might have served as a foundation for the origin of cells. The idea is that the vents created a natural version of the proton gradient across cell membranes and that the original biological membranes were assembled in this favourable environment.

The nature of the first cells is a subject of enthusiastic conjecture in the absence of objective data beyond the evident presence of life today. The relative timing of the origin of archaea and bacteria is unknown, but fossils show that filamentous prokaryotes resembling today's cyanobacteria had evolved at least 3.5 billion years ago. It is possible that these microbes performed oxygenic photosynthesis, but measurements of radioactive isotopes show that the sharp rise in atmospheric oxygen known as the Great Oxygenation Event (GOE) did not occur until 2.3 billion years ago. The putative billion-year discrepancy between oxygenic photosynthesis and the oxygenation of the atmosphere may be explained by the transformation of the planet's highly reduced marine chemistry to a more oxidized state. Oxygen released by photosynthesis would have been quenched by reacting with elements in their reduced forms for hundreds of millions of years after the evolution of cyanobacteria. Deposits of ferric iron called banded iron formations are found in sedimentary rocks of the same age as the GOE. These may reflect the tipping point from reductive to oxidative chemistry that allowed oxygen to accumulate in the atmosphere.

The origin of the eukaryotes has been linked to the GOE, but the appearance of cells with nuclei is very difficult to date and there are a number of competing ideas for the mechanism that produced them. Multiple lines of irrefutable evidence prove the bacterial origin of mitochondria and chloroplasts in eukaryotic cells. Both types of organelle contain single chromosomes typical of bacteria, they contain membranes that resemble the cell membranes of bacteria and perform some of the same kinds of biochemical reactions, and, thirdly, ribosomes that perform protein synthesis inside the mitochondria and chloroplasts resemble prokaryotic ribosomes. These observations support the endosymbiotic model for eukaryote evolution, which is one of the mainsprings of modern biology.

The origin of the nucleus that defines the eukaryotes is more controversial. Accumulating evidence supports the idea that the nucleus evolved in a specific group of archaea and that later fusion with a bacterium generated the ancestral mitochondrion. This model, called the eocyte hypothesis, is bolstered by the numerous commonalities between the genomes of archaea and eukaryotes. It recommends the concept of arranging all organisms into two principal domains, namely the bacteria and archaea, and treating the eukaryotes as a group *within* the archaea. This conflicts with the three-domain organization of life—bacteria, archaea, and eukaryotes—that has held sway among microbiologists for more than thirty years. Another suggestion is that the nucleus evolved *after* the absorption of a bacterium by an archaea. In this model, H_2 and CO_2 produced by fermentation in the bacterium supported the methanogenic metabolism ($H_2 + CO_2 \rightarrow CH_4$) of the host archaea. New uses of genomic data to test different models for eukaryote origins offer a marvellous illustration of the vitality of this modern scientific adventure.

Chapter 7
Microbes in agriculture and biotechnology

Natural mechanisms that control soil fertility are inadequate to the task of supporting modern intensive agriculture and a human population that will probably exceed nine billion in 2050. Soils exposed by deforestation lack the physical structure, chemical composition, and indigenous populations of microbes needed to cultivate consecutive harvests of cereals and other food crops. Intensive grazing of cattle on deforested land is similarly unsustainable. Even in areas where soils are exceedingly rich, crop productivity falls without soil amendment with organic matter and inorganic fertilizers. Today's global agrochemical business produces 2.4 billion tons of grains (more than 300 kilograms for every person) and 500 million tons of oilseeds. This industry relies on petroleum and natural gas that are used to drive farm machinery, power the production of fertilizers, and serve as the raw materials for the synthesis of pesticides. The cost of these agricultural practices and their contribution to climate change have stimulated applied microbiological research to improve soil fertility and combat plant diseases. Research on microbes has also been instrumental in the development of genetically modified (GM) food. These initiatives will be considered in this chapter before introducing some of the wider uses of microorganisms in biotechnology and their longstanding importance in baking, brewing, and the production of other foods.

Microbial effects on soil fertility include nitrogen fixation by *Rhizobium* in root nodules and by free-living bacteria, solubilization of phosphorus by bacteria and fungi, and breakdown of organic matter. The use of pure cultures of live microorganisms as 'biofertilizers' is a relatively new business, but the value of this market is projected to exceed $10 billion by 2017. Inoculation of soybean seeds with biofertilizers containing rhizobia has been practised for decades: this increases the number of nodules per plant and improves crop productivity. The addition of the free-living nitrogen-fixing bacterium *Azospirillum* to the seeds, or to the furrows in which they are planted, can increase crop productivity further. *Azospirillum*, *Azotobacter*, and other free-living bacteria are also used as biofertilizers for wheat and other crops that do not form root nodules. Filamentous cyanobacteria are applied to rice crops as free-living bacteria or as symbionts within the leaves of the aquatic fern *Azolla* that floats in the flooded paddy fields. Nitrogen absorbed by this relationship is released into the water as the plant tissues decompose. Mycorrhizal fungi that supply plants with a range of minerals are marketed as root treatments for trees and shrubs, and non-mycorrhizal soil fungi that increase phosphorus availability have also been commercialized as biofertilizers. The compatibility of biofertilizers with organic farming is a significant attribute of this developing technology. With only 1 per cent of the world's farmland cultivated using these methods, however, our reliance on fossils fuels for food production is unlikely to diminish soon.

Pathogenic fungi, bacteria, and viruses destroy a significant proportion of crops during cultivation and saprotrophs damage agricultural products during storage. Microorganisms are used, with varying success, as agents of biological control to limit crop losses caused by insect and nematode pests. These 'biopesticides' include baculoviruses, soil bacteria, and fungi. Baculoviruses are double-stranded DNA viruses that infect insects. Multiple infectious baculovirus particles are formed inside protein matrices called occlusion bodies that protect the virus when it is sprayed on the surface of vegetation. When the occlusion bodies are eaten by insect larvae, the individual virions are released in the gut and infect the host cells. Replication of the virus produces so much tissue damage that the insects disintegrate, releasing more infectious particles that cause the next round of infection. Baculoviruses have been used to control codling moth infestations of apples and pears, and are sprayed over thousands of acres of forest in the United States every year to kill gypsy moths.

Strains of *Bacillus*, *Pseudomonas*, and *Streptomyces* are effective at promoting plant growth and prevent certain diseases by outcompeting pathogenic microorganisms that otherwise proliferate in the rhizosphere. *Bacillus thuringiensis* is a particularly effective agent that produces a protein called Bt toxin that kills insect larvae by destroying the insect gut. This bacterium has been used as the active ingredient of insecticides for decades. The fungus *Beauveria bassiana* is another effective biopesticide that is used to control aphids, and the yeast *Candida oleophila* is applied to the surface of fruits to prevent post-harvest rotting by filamentous fungi.

The ubiquity of genetically modified (GM) plants in our food supply, and the controversial nature of these commodities, has obscured the revolutionary technology involved in their production. *Agrobacterium tumifaciens* is a relative of the nitrogen-fixing rhizobia that form nodules on the roots of legumes. It is not a mutualist, however, and infects plants through wounds, transmitting a plasmid whose tumour-inducing genes become integrated into the host genome. This natural disease process has been exploited in plant biotechnology and the bacterium is used as a vector to introduce foreign genes into tomato, potato, soybean, and fruit trees. Genetic transformation of plants is also accomplished using gene guns to blast gold particles coated with DNA into plant tissues. Both methods are used to modify crops with the Bt toxin gene from *Bacillus thuringiensis*, allowing the plant to produce its own insecticide rather than spraying the crop with biopesticide containing live bacteria.

While there is resistance to the cultivation of GM foods in Europe, the Bt toxin gene is incorporated into more than 75 per cent of the corn (maize) and cotton planted in the United States. Complications associated with the introduction of Bt toxins include the evolution of Bt-resistant insect pests, apparent alterations to the soil microbiome, and uncertainties about the potential for damage to non-target insect species. The possibility of Bt accumulation in soils is also a concern because this might damage soil invertebrates. One of the potential advantages of the technology is that farmers planting GM crops would be able to use smaller quantities of pesticides, but some studies indicate that this has not happened. Much of the current data that might help adjudicate these issues is equivocal.

The use of antibiotics for increasing the yields in meat and dairy production is another controversial issue in agricultural microbiology. Farming accounts for half of all antibiotic use and much of this is consumed in raising livestock in intensive farming. Cows, pigs, and poultry receive antibiotics in their feed as a disease preventive rather than treatment for any active infection. The fact that antibiotics promote animal growth and reduce the populations of harmful bacteria on meat products are powerful incentives for their continued routine use. The potential problem with this habit, along with the overprescription of antibiotics by physicians, is that it may accelerate the development of antibiotic-resistant strains of *Salmonella*, *Staphylococcus*, *Escherichia coli*, and other bacteria. Antibiotic addition to animal feed is banned in the European Union and future legislation may force American farmers to follow suit. Reports on the global potential for the development of antibiotic-resistant bacteria highlight the huge scale of pig farming in China and poorly regulated use of antibiotics in this industry.

There is considerable irony in our relationship with antibiotics. Humans have been afflicted by infectious diseases throughout history and the treatment of these illnesses using antibiotics is one of the signal victories of modern medicine. On the downside, the overuse of these compounds has spawned an unprecedented threat to civilization. It would be unreasonable to suggest that the cure is worse than the illness, but this may be the judgement of future generations threatened by a pernicious microbiology that is impervious to our medications.

There is little ambiguity when we consider the impact of other pharmaceutical products synthesized by microorganisms. Genetically engineered bacteria produce insulin and other human hormones, proteins that promote blood clotting and others that dissolve clots, enzymes used for treating the symptoms of cystic fibrosis, and agents used to treat multiple sclerosis, viral infections, and certain cancers. To produce these human proteins in bacteria, genetic engineers have to reformat the human gene so that it is compatible with protein synthesis in the bacterial host. The problem with human genes, and the genes of all eukaryotes, is that they are interrupted with non-coding sequences called introns (Chapter 3). When genes are transcribed in eukaryotic cells, the sequences derived from introns are removed (spliced) from the mature forms of the messenger RNA (mRNA) molecules before they are translated into proteins. Because bacteria lack this splicing machinery, the introns must be removed from the engineered genes. This is done by using mature mRNA from human cells to create a DNA copy without introns, or by synthesizing the DNA from scratch based upon the known sequence of the gene. In either case, the DNA, referred to as complementary DNA (cDNA), is inserted into a plasmid that is incorporated into the bacterium.

'Live' viruses have been used in a number of experimental gene therapies for serious illnesses caused by mutations in the human genome. Two methods have been investigated. The first involves infection of the patient with the virus that operates as a vector that carries the corrected gene into each cell that it infects. In the second method, cells are isolated from the patient, transformed using a virus or plasmid, cultured in the laboratory, and reintroduced to the patient. Retroviruses and adenoviruses have been chosen for most experiments. Retroviruses will insert genes into the patient's chromosomes; adenoviruses introduce genes into the nucleus of the host cell without integrating foreign genetic material into the chromosomes. These treatments are very controversial and early trials in which viruses were used to insert corrected copies of genes into patients were considered unsuccessful. Problems with viral gene therapies include immune reactions to the viral vectors, viral infection of the wrong cells, and the induction of tumours caused by disruption of the host genome. The technology is improving rapidly, however, and recent investigations using viral therapy to treat thalassemia, haemophilia, certain forms of leukaemia, and autoimmune diseases seem promising.

Escherichia coli is the commonest microbe used to synthesize human proteins, but engineered strains of yeast, *Saccharomyces cerevisiae*, are also very valuable. Being a eukaryote, yeast possesses the splicing mechanism needed to express human genes without prior removal of introns. Other bacteria and fungi are also used in genetic engineering. Production of proteins by recombinant microorganisms is carried out in stirred tank fermenters in which sterile growth medium is inoculated, aerated, stirred, monitored by sensing instruments, heated or cooled, and fed with additional materials. Fermentation tanks or bioreactors are manufactured from stainless steel to avoid corrosion and leaching of toxic metals into the medium. Batch culture and continuous culture methods are used to produce insulin and other drugs. In batch culture, the level of nutrients declines as the density of cells increases and the process is stopped to harvest the product. Continuous culture allows the operator to maintain optimal conditions for fermentation for many weeks by programming cycles of nutrient injection. Fermentation products can be harvested repeatedly or continuously using this method. (In addition to the use of fermentation to refer to the anaerobic conversion of sugars into ethanol, the term is used more broadly to describe any of the 'industrial' transformations carried out by microorganisms.) Bioreactors that maintain optimal growth conditions for the cultured microorganisms by the continuous removal of the spent culture liquid and addition of fresh medium are called chemostats.

Fermentation by yeasts and bacteria produces carbon dioxide and ethanol under conditions of low oxygen availability or high concentrations of sugar. It allows these organisms to sustain a low level of energy production—relative to aerobic as well as anaerobic respiration—through the reactions of glycolysis (Chapter 2). This metabolic mechanism is the basis for wine and beer production. *Saccharomyces cerevisiae* is used for wine and beer fermentation and related yeasts are used for making lager and cider. The growth of most bacteria is avoided in winemaking and beer brewing because their proliferation tends to spoil the flavour of the drink. *Oenococcus oeni*, a firmicute bacterium, is a welcome exception that reduces the acidity of wine by fermenting malic acid, and some Belgian beers develop their prized sourness from the growth of *Lactobacillus*.

Wine and cider are made from sugary plant juices and beer is brewed from starchy plant materials whose starch is converted to sugars before fermentation to produce alcohol. Red wines are fermented from grapes with red or purple/black skins whose phenolic pigments colour the wine. Grape skins and seeds are included in the fermentation of red wines. White wines are made from juice expressed from grapes without using the skins. Most white wines are produced from white grapes, but dark grapes are used for a few white wines because the pigments separate with the skin. Crushing of harvested grapes releases juice and this is fermented in barrels, open vats, or in industrial fermenters. Fermentation in traditional winemaking is carried out by wild yeasts that are transferred to the grape juice, or must, from equipment in the winery or from the surrounding environment. Modern vintners are more likely to inoculate their grape juices with specific yeast strains. *Saccharomyces* grows until the alcohol content reaches 10–12 per cent and the residual concentration of sugars defines the level of sweetness in the wine. After this first phase of the fermentation the wine is transferred to casks for maturation and storage.

In malting, the first stage of beer brewing, barley grains are steeped in water and allowed to germinate. After enzymes in the grains break down proteins and starch to form amino acids and sugars, the germinating barley is dried with a stream of hot air in a kiln to produce malt. After kilning, the malt is milled into a course flour which is mixed with warm water to form the mash. In the last non-microbial step in the brewing process, solids are separated from the mash to leave a sugary liquid wort that is fermented by yeasts. Dried female flowers of the hop plant are added to the wort and the mixture is boiled to denature the enzymes and add flavour. Residues from the hops are removed, the cooled mixture is aerated to promote yeast growth, and fermentation is carried out in open tanks or closed steel fermenters. For brewing ales, top-fermenting strains of yeast mix throughout the wort and accumulate in foam that collects at the surface. Bottom yeasts used in lager brewing settle at the bottom of the vat. Oxygen levels fall during both kinds of fermentation, and the yeast converts sugars to alcohol under anaerobic conditions with limited increase in biomass.

Saccharomyces cerevisiae is also used to ferment dough to produce bread and other baked foods. Romans used top yeast harvested from breweries for bread making and this practice continued through the 19th century. Concentrated forms of single yeast strains are manufactured for today's bakeries and for baking at home. These are prepared by fermenting molasses under aerated conditions to maximize respiration and biomass accumulation. (The opposite of brewing.) Synthesis of the sugar alcohol trehalose is stimulated under these conditions and this compound protects the yeast cells when they are freeze-dried. A concentrated liquid form of yeast, called cream yeast, is used by large bakeries. Dough is made by mixing water, yeast, and salt; amylases in the dough break down starch to form sugars which are fermented by the yeast; other ingredients can be added before the dough is mixed (kneaded), and the fermentation proceeds with the release of carbon dioxide which causes the dough to rise. The yeast is killed by ethanol produced during the fermentation and this evaporates when the bread is baked.

In cheese production, bacteria ferment lactose in milk to form lactic acid and impart many of the flavours characteristic of different varieties. Solid curds and liquid whey are separated using the enzyme rennin to coagulate the milk. Rennin was collected from the stomach lining of slaughtered calves but most of today's producers use an enzyme produced by recombinant bacteria and fungi. Yeasts and filamentous fungi play subsidiary roles in flavouring and ripening cheeses. The white rinds of Brie and Camembert cheeses are formed by dense colonies of filamentous hyphae of the ascomycete *Penicillium camemberti*. Bacteria create the cavities in blue-veined cheeses and these are colonized by *Penicillium roqueforti* that produces spores and imparts glorious flavours to Roquefort, Gorgonzola, Stilton, and Danish Blue.

Many Asian foods and drinks are produced by fungal fermentations. Tempe, which is a traditional Javanese food, is prepared by fermenting beans and cereals with *Rhizopus* and *Mucor* (filamentous zygomycete fungi). Zygomycete fungi are also used to ferment blocks of tofu (soybean curd) to make the cheese-like Furu or Sufu that is popular in China. Soy sauce is a more familiar soybean product whose complex manufacture involves a five-day fermentation by *Aspergillus oryzae* and *Aspergillus sojae* followed by a second stage, year-long fermentation by yeasts and lactic acid.

Fungi produce ethanol for biofuels by fermenting sugars extracted from sugarcane, corn, and other crops. In Brazil, juice is extracted from sugarcane and concentrated to produce a sucrose-rich syrup, or molasses, which is fermented by yeast. Fibrous material remaining after juice extraction is burned as a fuel for this industrial process. Corn ethanol is a less efficient form of biofuel produced in the United States. Because corn is rich in starch, corn kernels are processed using purified enzymes to release fermentable sugars. Second-generation biofuel production from rice straw and other plant materials containing lignin and cellulose is an emerging technology. Cyanobacteria, eukaryotic green algae, and diatoms are also under investigation as potential sources of biofuel. The rationale for using photosynthetic microorganisms for fuel generation is that a high proportion of the biomass of these cells is stored as oils that can be refined into diesel or jet fuel.

The biochemical versatility of microorganisms is utilized for releasing metals from ores in mining operations and is also valuable for cleaning and detoxifying environments damaged by our industries. Metal extraction by bioleaching is used to recover metals from piles of ore called 'leach dumps'. The iron-oxidizing proteobacterium *Acidithiobacillus ferrooxidans* is the workhorse of this industry. In the recovery of copper from piles of low-grade copper ore containing copper sulfides, ferric iron (Fe^{3+}) reacts with CuS (copper monosulfide or covellite) to release soluble copper ions (Cu^{2+}). The dissolved copper is precipitated in ponds by reacting with iron supplied as shredded scrap metal. Ferrous iron (Fe^{2+}) is produced from this second reaction. Slurry from the precipitation ponds is pumped into a second set of ponds in which *Acidithiobacillus* regenerates ferric iron by oxidizing Fe^{2+} to Fe^{3+}. It uses electrons stripped from Fe^{2+} to fix carbon dioxide. Water enriched with ferric iron is pumped back to the top of the ore pile to sustain the solubilization of copper. Management of these kinds of redox reactions is used to leach zinc, nickel, gold, uranium, and other metals from ores, and has been used to recover materials from dust generated by recycling computer circuit boards. *Acidithiobacillus* is part of a community of bacteria that grows in oxidation ponds.

There is a lot of interest in exploiting the biochemical activities of bacteria and fungi to detoxify soil and water contaminated with arsenic and other metals, to remediate environments polluted with hydrocarbons by the oil and gas industries, and to decompose synthetic chemicals that pose a hazard to human health. This research subject is called bioremediation. Mycorrhizal fungi have proven effectiveness at accumulating toxic metals, including radioactive elements from soil. One way to utilize the absorptive capacity of large fungal colonies is to plant trees in contaminated sites and allow them to recruit mycorrhizal partners from the existing soil microbiome. Alternatively, the roots of saplings can be inoculated with specific fungi before planting.

Phanerochaete species are white rot fungi that secrete peroxidases and laccases that decompose the lignin in diseased trees and fallen wood. The same enzymes are active against aromatic pollutants including petroleum products, chlorinated chemicals from wood preservatives, halogenated compounds used as flame retardants, and explosives. Pesticides and other agrochemicals are also targets for the enzymes secreted by white rot fungi. Encouraging results have come from experiments in which the fungi are grown on sawdust and woodchips impregnated with these contaminants, but large-scale studies have not been published. Similar optimism is expressed by microbiologists studying fungi and bacteria capable of biodegrading toluene, naphthalene, and other highly toxic organic compounds, but the practicality of these bioremediation strategies is uncertain.

Oil spills from tankers and drilling rigs have produced some of the most widespread cases of environmental pollution, damaging productive fisheries, killing marine mammals and birds, and wrecking local economies. There is some capacity among marine bacteria to biodegrade oil droplets and it may be possible to stimulate the activity of these microorganisms to aid cleanup operations. *Alcanivorax borkumensis* is a marine proteobacterium that uses an array of enzymes to oxidize saturated hydrocarbons or alkanes. The bacterium aids its digestive process by releasing surfactant molecules that help break up the oil into small droplets. The bacterium is aerobic, making it most effective at oil treatment in the surface of the ocean. After an oil spill, low molecular weight alkanes evaporate into the air and the larger molecules sink at various rates. Heavier hydrocarbons form plumes that sink to the seafloor and these components of the oil cannot be broken down by aerobic bacteria. *Alcanivorax* populations increase rapidly after an oil spill, forming a bacterial bloom in the polluted water. Growth of this bacterium is limited by the availability of phosphorus and nitrogen, and the addition of these elements to 'fertilize' the bacteria is one of the strategies considered for accelerating oil cleanup. Another hydrocarbon-consuming bacterium, *Oleispira antarctica*, was isolated from coastal water in the Antarctic and is viewed as a potential bioremediation agent for oil spills in cold water environments.

Pollution of the biosphere in ways that harm humans and other organisms is an undeniable corollary of our industrial and agricultural activities. The relationship between fossil fuel consumption and climate change is an even more pressing concern from the point of view of Earth's long-term habitability. By detoxifying soil and water and absorbing carbon dioxide, microorganisms operate as formidable buffers against our mounting population and its degradation of natural resources. There are, however, limits to the capacity of bacteria, archaea, and eukaryotic microbes to maintain chemical homeostasis in the biosphere. We will know, at least in hindsight, when the threshold for microbial purification of soil, water, and the air is exceeded in a manner that renders our future unsupportable.

In the meantime, our appreciation of the microbial dominion over the biosphere is deepening. The metagenomic exploration of familiar terrestrial and aquatic habitats is showing fantastic numbers of microorganisms, including the surprising preponderance of viruses everywhere we look for them. We have identified hidden 'worlds' of bacteria in deep sediments and thriving in other environments once considered too extreme for any living thing. And, focusing on ourselves, recent discoveries about the human microbiome offer a fresh and inspiring view of *Homo sapiens* as a multifaceted ecosystem—more bacterial in cell numbers than we are animal. Each of us is a participant in a magnificent symbiosis, born into microbiology and digested by it at the end. The microbes are everywhere and will outlive us by an eternity.

Further reading

Microbiology is such a vibrant area of research that any book that aims to be current is certain to miss important findings within months of publication. Academic journals are the best source of current information and the following periodicals offer reviews of ongoing microbiological research: *Annual Review of Microbiology*, *Clinical Microbiology Reviews*, *Current Opinion in Microbiology*, *Current Topics in Microbiology and Immunology*, *FEMS Microbiology Letters*, *FEMS Microbiology Reviews*, *Frontiers in Microbiology*, *Fungal Biology Reviews*, *Letters in Applied Microbiology*, *Microbiology and Molecular Biology Reviews*, *Microbiology Today*, *Nature Reviews Microbiology*, and *Trends in Microbiology*.

Chapter 1: Microbial diversity

There are several introductory textbooks on microbiology.
M. T. Madigan, J. M. Martinko, D. Stahl, and D. P. Clark, *Brock Biology of Microorganisms*, 13th edition (San Francisco: Benjamin Cummins, 2010) is among the best, providing a wealth of information supported with excellent illustrations. It is particularly strong on the complicated topic of microbial metabolism. This book, like most microbiology texts, concentrates on prokaryotes and viruses. One of my previous books provides a contemporary view of biodiversity with emphasis on eukaryotic microbes: N. P. Money, *The Amoeba in the Room: Lives of the Microbes* (Oxford University Press, 2014). Photosynthetic eukaryotic microorganisms, or algae, are introduced in greater detail in L. Barsanti and P. Gualtieri, *Algae: Anatomy, Biochemistry, and Biotechnology*, 2nd edition (Boca Raton, Fla.: CRC

Press, 2014). The following website is a useful supplement to these books because it presents a clickable evolutionary tree that directs readers to details on individual groups of microorganisms: <http://www.tolweb.org/tree/phylogeny.html>.

Chapter 2: How microbes operate

F. M. Harold, *The Vital Force: A Study of Bioenergetics* (New York: W. H. Freeman, 1986).

D. G. Nicholls and S. Ferguson, *Bioenergetics*, 4th edition (Amsterdam: Elsevier, Academic Press, 2013).

Chapter 3: Microbial genetics and molecular microbiology

B. Alberts et al., *Molecular Biology of the Cell*, 5th edition (New York: Garland Science, 2007).

L. Snyder, J. E. Peters, T. M. Henkin, and W. Champness, *Molecular Genetics of Bacteria*, 4th edition (Washington, DC: ASM Press, 2013).

Chapter 4: Viruses

J. Carter and V. Saunders, *Virology: Principles and Applications*, 2nd edition (Chichester: Wiley: 2013).

D. H. Crawford, *Viruses: A Very Short Introduction* (Oxford University Press, 2011).

C. Zimmer, *A Planet of Viruses* (University of Chicago Press, 2011).

Chapter 5: Microbiology of human health and disease

S. G. B. Amyes, *Bacteria: A Very Short Introduction* (Oxford University Press, 2011).

M. J. Blaser, *Missing Microbes: How the Overuse of Antibiotics is Fueling our Modern Plagues* (New York: Henry Holt, 2014).

L. Collier, J. Oxford, and P. Kellam, *Human Virology*, 4th edition (Oxford University Press, 2011).

K. Murphy, *Janeway's Immunobiology*, 8th edition (New York: Garland, 2011).

L. M. Sompayrac, *How the Immune System Works*, 4th edition (Hoboken, NJ: Wiley-Blackwell, 2012).

Chapter 6: Microbial ecology and evolution

C. Gerday and N. Glansdorff (eds.), *Physiology and Biochemistry of Extremophiles* (Washington, DC: ASM Press, 2007).

D. L. Kirchman (ed.), *Microbial Ecology of the Oceans*, 2nd edition (Hoboken, NJ: Wiley-Blackwell, 2008).

D. L. Kirchman, *Processes in Microbial Ecology* (Oxford University Press, 2012).

R. V. Miller and L. G. White, *Polar Microbiology: Life in a Deep Freeze* (Washington, DC: ASM Press, 2012).

F. Rohwer, M. Youle, and D. Vosten, *Coral Reefs in the Microbial Seas* (Basalt, Colo.: Plaid Press, 2010).

Chapter 7: Microbes in agriculture and biotechnology

M. P. Doyle and R. L. Buchanan, *Food Microbiology: Fundamentals and Frontiers*, 4th edition (Washington, DC: ASM Press, 2013).

B. R. Glick, J. J. Pasternak, and C. L. Patten, *Molecular Biotechnology: Principles and Applications of Recombinant DNA*, 4th edition (Washington, DC: ASM Press, 2009).

E. A. Paul, *Soil Microbiology, Ecology and Agriculture*, 3rd edition (Amsterdam: Elsevier, Academic Press, 2007).

Index

Index

Index

ONLINE CATALOGUE
A Very Short Introduction

Our online catalogue is designed to make it easy to find your ideal Very Short Introduction. View the entire collection by subject area, watch author videos, read sample chapters, and download reading guides.

http://fds.oup.com/www.oup.co.uk/general/vsi/index.html

SOCIAL MEDIA
Very Short Introduction

Join our community
www.oup.com/vsi

- Join us online at the official Very Short Introductions **Facebook** page.
- Access the thoughts and musings of our authors with our online **blog**.
- Sign up for our monthly **e-newsletter** to receive information on all new titles publishing that month.
- Browse the full range of Very Short Introductions online.
- Read **extracts** from the Introductions for free.
- Visit our library of **Reading Guides**. These guides, written by our expert authors will help you to question again, why you think what you think.
- If you are a teacher or lecturer you can order inspection copies quickly and simply via our website.